SILICON DOCKS

First published in 2015 by
Liberties Press
140 Terenure Road North | Terenure | Dublin 6W
T: +353 (1) 9056073 | www.libertiespress.com | E: info@libertiespress.com

Trade enquiries to Gill & Macmillan Distribution
Hume Avenue | Park West | Dublin 12
T: +353 (1) 500 9534 | F: +353 (1) 500 9595 | E: sales@gillmacmillan.ie

Distributed in the United Kingdom by
Turnaround Publisher Services
Unit 3 | Olympia Trading Estate | Coburg Road | London N22 6TZ
T: +44 (0) 20 8829 3000 | E: orders@turnaround-uk.com

Distributed in the United States by
International Publishers Marketing
22841 Quicksilver Dr | Dulles, VA 20166
T: +1 (703) 661-1586 | F: +1 (703) 661-1547 | E: ipmmail@presswarehouse.com

ISBN: 978-1-909718-87-6
2 4 6 8 10 9 7 5 3 1

A CIP record for this title is available from the British Library.

Cover design by Karen Vaughan – Liberties Press
Cover image © Conor Nolan
Internal design by Liberties Press

SILICON DOCKS

The Rise of Dublin as a Global Tech Hub

Edited by
Pamela Newenham

To my parents, Rowland and Maggie, and sister, Janet

Table of Contents

Foreword

Barry O'Sullivan

On a grey, misty morning in December 2010, I landed at Dublin airport after a long flight from San Francisco via Heathrow. I was feeling great – the division I was running at Cisco was booming, and I was looking forward to seeing my family – just a few meetings in Dublin, then on to Galway for Christmas.

The chat with the taxi driver on the way into town changed my mood. This was just months after the national humiliation of the EU/IMF bailout, and it was clear that people were reeling from the shock and scale of the economic contraction. He talked about the fall-off in his business and the struggles with a huge mortgage. I'll never forget his sad summary: 'No Christmas in our house this year.'

My first meeting was at the IFSC with a fund manager, regarding a small investment fund I was putting together with some California-based technology investors. It was clear from the start that they would not be participating. 'Look,' the fund manager said, 'it's hopeless. We're not doing anything in Ireland. Nothing good is happening here.'

I was disappointed but not discouraged, because I knew he was wrong. As a founder of the Irish Technology Leadership Group, I had met some incredible Irish technology companies over the previous few years, as we hosted them at various events in Silicon Valley. People like Connor Murphy of Datahug and Pat Phelan of Trustev had an infectious enthusiasm and fearlessness that made me believe the future for Ireland was far from hopeless.

So, with the IFSC behind me, I walked across Samuel Beckett Bridge to my meeting with a small start-up on Barrow Street, and it struck me that I was leaving behind the tired old order that had failed Ireland. Across the bridge in front of me was the future, full of hope and possibility – the shiny new collection of buildings housing global technology companies, start-ups and venture capitalists, now known as Silicon Docks.

This book, by some of Ireland's leading business and technology journalists, comes at an important moment because, in many ways, the technology industry in Ireland is at a crossroads. The authors chart the history of Silicon Docks, from the establishment of the Dublin Docklands Development Authority in 1997, to the arrival of Google in 2004, the subsequent addition of social media giants Facebook and Twitter, and the beginnings of a start-up revolution that is promising but still in its early stages.

The success of Silicon Docks is a combination of visionary public policy and the blind luck of good timing. The early years of the project coincided with a global shift for technology companies – from being located in suburban technology parks (think IBM in Blanchardstown or Microsoft in Sandyford) to cool new 'innovation districts' in city centres. Such innovation districts have sprung up in major cities such as New York, London and Barcelona. The new generation of young tech workers want to live and work in cities and bike, walk or take public transport to work. The most striking example is San Francisco, where start-ups are forsaking the suburbs of Sunnyvale and San Jose for the cool lofts and warehouses south of Market Street in the city.

Bruce Katz and Julie Wagner of The Brookings Institution have led research on the new global geographical phenomenon of innovation districts, which they define as:

> geographic areas where leading-edge anchor institutions and companies cluster and connect with start-ups, business incubators and

> accelerators . . . physically compact, transit-accessible, and technically-wired and offer mixed-use housing, office, and retail.

This definition could have been written to describe Silicon Docks. On the doorstep of Trinity College, the area includes not only global tech companies and local start-ups, but also venture capitalists and the accelerators Dogpatch and Wayra.

What does long term success look like for Silicon Docks? I believe we will have achieved it when the biggest employers are indigenous Irish tech companies that have achieved global success. We have most of the ingredients to make this happen, but there are three issues that need to be addressed first and they are related to the ingredients that fuel successful start-ups: talent and money.

1. *Low R&D investment by some multinationals.* In the technology industry, the most important talent is engineering; people who can write software and build products. Co-locating big companies with start-ups sounds great, but the idea that there is a fungible engineering talent pool across the multinationals and start-ups in Silicon Docks doesn't stand up to scrutiny. This is for the simple reason that Facebook, Google, Twitter and LinkedIn do not build their core software products in Dublin. These are designed and coded in Menlo Park, Mountain View, San Francisco and, to some extent, London. There is no pool of software-engineering talent coming out of their Dublin offices. What's more, the lack of R&D in these operations means that they are not firmly anchored in Dublin. We should set a strategic goal to secure significant R&D investment by these companies in Dublin over the next couple of years.

2. *Irish venture funds need to be replenished.* Venture funds usually invest in five-year cycles, and the Irish funds are coming to an end. Put more simply, they are running out of money. A structural change to the Irish pension industry (the move to defined-contribution plans) has made pension funds more conservative and reluctant to make higher-risk, higher-return venture-class investments. Two-thirds of all new jobs are now created by start-ups. This job-creating machine will grind to a halt unless Irish funds can raise money. Given this strategic imperative, one possible solution is the allocate some of the Ireland Strategic Investment Fund (formerly the National Pensions Reserve Fund) to venture capital for Irish start-ups.
3. *Ireland has an uncompetitive tax regime.* This sounds counter-intuitive given that we have one of the most attractive corporation tax rates globally. The problem is that we also have amongst the most uncompetitive income tax and capital gains tax (CGT) regimes globally. I know from personal experience in my company that it's virtually impossible to attract highly paid technical talent to Ireland when they hear our top rate of tax is 52 percent. It's a deal killer and it needs to change.

Similarly, the value-creation model for technology start-ups is focused on capital appreciation. Profits in most cases are reinvested in the company to drive higher shareholder value. So a competitive capital gains tax rate is critical. It is not sustainable for Irish founders to have a CGT rate of 33 percent while entrepreneurs in Northern Ireland can access a 10-percent rate.

In this book, the authors lay out the history, achievements and opportunities of Silicon Docks. The success of the Silicon Docks project is testament to a unique Irish combination of forward thinking public policy, risk-taking and hard work. There is now an opportunity for us to lead the world from a small enclave on the banks of the River Liffey. I believe this new, young generation of entrepreneurs has the one quality required to get us there – determination.

As I think back to my walk over the Samuel Beckett Bridge on that rainy day in 2010, I am reminded of his famous quote, which should be an encouragement to anyone who wants to overcome setbacks and succeed: 'Ever tried. Ever failed. No matter. Try Again. Fail again. Fail better.'

—Barry O'Sullivan
Palo Alto, California
December 2014

Preface

Pamela Newenham

There are numerous tech companies, start-ups, entrepreneurs, incubators and accelerator programmes in Silicon Docks, the rest of Dublin and throughout Ireland that are worthy of mention in this book. However, due to space considerations, many are omitted. Even focussing on the Grand Canal Dock/Silicon Docks area, it was very difficult to keep up with the fast-growing tech scene. More companies were arriving from abroad and new start-ups were emerging every week.

The start-up scene, tech scene and tax environment were also changing rapidly. The book wasn't long filed when the European Commission published preliminary findings of its investigation into Apple's tax affairs in Ireland. The Commission said two tax deals agreed between Ireland and the US computer giant amounted to illegal state aid. Two weeks earlier, the OECD published proposals aimed at putting a stop to companies such as Apple shifting their profits into overseas tax havens. Then, in October, the government closed the 'double-Irish' tax loophole, used to move profits overseas. At the same time, changes were happening in the tech world. LinkedIn acquired a site in Dublin on which to build a new international headquarters, serial entrepreneur Niamh Bushnell was appointed Dublin's first commissioner for start-ups, and tech companies such as Groupon made further job announcements.

We have tried to stay on top of everything, adding in all major changes. However, it is likely that further tech-related announcements and changes will happen between the last filing date and publication of this book.

Numerous people helped with this book, but several went above on beyond when it came to research, securing interviews, taking pictures and digging up old photos. They are Emmet Oliver of IDA Ireland, Lorcan O'Sullivan of Enterprise Ireland, Celine Crawford of MKC Communications, Mark O'Toole of the Web Summit and Kristina Petersone.

Three people were a tremendous help when it came to proofreading the book. They were my parents Rowland and Maggie, and my friend Sharon Pennick, who spent nearly her entire holiday in Mexico proofreading. My friend Jasmine Godwin was also a great help, ensuring I was well-fed throughout the writing-and-editing process.

I would like to thank the team at Liberties Press, especially Sam Tranum, Karen Vaughan, Ailish White and Sean O'Keeffe, for being patient when deadlines weren't met, and putting up with my fussiness over images, the cover and the title. I would also like to thank everyone in the *Irish Times* business section for all their support, and for affording me time off to work on this book.

Finally, I would like to thank my *Irish Times* colleague Fiona Reddan for recommending me to Liberties Press.

—Pamela Newenham
December 2014

Introduction

Pamela Newenham

In 1939, William Hewlett and David Packard established a little electronics company in Palo Alto, a small city southeast of San Francisco. That company would later become multinational computer giant Hewlett Packard, and the garage in which they created it would be dubbed 'the birthplace of Silicon Valley'.

Almost sixty years later, in another garage not far away, Sergey Brin and Larry Page founded Internet-search firm Google. That company would later become the founding father of Silicon Docks.

Establishing its European headquarters on Barrow Street in 2004, Google would lead a long line of Internet companies to Grand Canal Dock. Located on the south side of Dublin's River Liffey, the area was once industrial wasteland, and a blot on the capital's cityscape. Today, it is Ireland's tech mecca, teaming with entrepreneurs, start-ups, incubators, and the European headquarters of companies such as Facebook, Squarespace, Riot Games, and Engine Yard, not to mention Irish tech firms such as Realex Payments and Mobile Travel Technologies.

Creating a Legal Infrastructure for High-Tech Companies

Situated in the southern part of the San Francisco Bay Area, Silicon Valley was first named so in the early 1970s, after the region's large number of silicon-chip manufacturers. However, it wasn't until the 1980s, when IBM introduced personal computers to the consumer market, that the moniker caught on.

Along with HP and IBM, Bell Labs, Fairchild Semiconductor, Adobe, Intel and Symantec were all in Silicon Valley at that stage. By the 1990s, the area would be home to one-third of the largest technology companies in the US.

The rise of Silicon Valley as a global centre for technology and innovation was strengthened by the legal environment of the time, as California state law precluded companies from putting non-compete clauses in employment contracts. This facilitated the movement of technology and ideas among workers, start-ups, and large firms.

The growth was also helped by the arrival of national and international law firms, which began setting up offices in the cities of Palo Alto, Menlo Park, and Mountain View from the 1980s onward. In 1996, Irish firm Matheson Ormsby Prentice (later renamed Matheson), became the first European law firm to open an office in Silicon Valley.

The law firms ensured the appropriate legal infrastructure was in place to support huge funding rounds, billion-dollar acquisitions, initial public offerings, and an increasing amount of intellectual property matters.

Like its US namesake, which was also a geographically ill-defined zone before it became a tech hub, Silicon Docks has attracted some of the country's top law firms, with Mason, Hayes and Curran, Dillon Eustace, Beauchamps, McCann Fitzgerald and Matheson all taking up residence in the area.

Silicon Ireland

The earliest mention of Ireland in the same sphere as Silicon

Valley was in 1995, when Harvard business professor Rosabeth Moss Kanter referred to the country as 'Silicon Bog'. In an interview with the *San Jose Mercury News*, Kanter said: 'Silicon Valley no longer is unique and no longer is able to monopolise the development of new technology. Everyone is already talking about Silicon Gulch in Austin, Silicon Mountain in Colorado, Silicon Forest in Seattle, Silicon Bog in Ireland, and Silicon Glen in Scotland'.[1]

Three years later, American technology magazine *Wired* referred to Ireland as 'Silicon Isle', noting that the country was the second-largest software exporter on the planet, with 'most major IT companies present and accounted for'.[2] At that stage, Apple, Microsoft, Hewlett Packard, EMC, Intel, IBM, Analog Devices, Oracle, and AOL all had operations in Ireland.

It wasn't until the twenty-first century, however, that the name 'Silicon Docks' would arise.

Early Days of the Docklands

For more than two hundred years, the Grand Canal Dock area was a thriving port, with boatbuilding, rope making, glass manufacturing, flour milling and gas production among the established industries. However, the introduction of shipping containers in 1956 led to a dramatic loss of employment in the docklands, with dockers no longer needed to load and unload cargo from the ships. This, along with a shift in manufacturing and distribution companies to business parks on the outskirts of the city, resulted in the physical and economic decline of the area. Schools and shops closed and, by 1996, the docklands population had halved.[3]

That same year, as part of the Irish government's budget, Minister for Finance Ruairi Quinn announced that the docklands were to be redeveloped. A docklands task force was established and, within two months, it had submitted a comprehensive report to the government, recommending the drafting of a master plan and the creation of a new authority to implement that plan. Debating the Dublin Docklands Development bill in Dáil Éireann on 27 February 1997, Senator Pat Magner said: 'We are not reinventing the wheel with this legislation. However, this will be the most exciting project in our lifetimes.'[4] He would later be proved right.

The government accepted the recommendations and, within months, the Dublin Docklands Development Act was signed into law, creating an authority tasked with the physical and social regeneration of Grand Canal Dock. The new authority got to work quickly, acquiring the derelict former gasworks site on Hanover Quay in 1997 and overseeing its decontamination.

The Game-changer

In the years that followed, Grand Canal Dock became one of the most intensively developed areas in the country, with construction of more than 73,000 square metres of office and residential space advanced in 2004.

It was also in 2004 that Google made the momentous decision to locate in the area, renting office space on Barrow Street. Unlike the multinational technology firms that had come to Ireland before it, the search giant wanted a site in the city, somewhere within easy reach of its future employees.

Google started out small, with the expectation that a few hundred jobs would be created. However, with more than 2,500 employees and several buildings on the street, Google is anything but small a decade later. Having spent more than €250 million purchasing four office buildings in the area, Google is the linchpin of the Silicon Docks tech cluster that developed.

2007 Onwards

Grand Canal Square, billed as an 'exciting new urban space' was opened in June 2007. Located between Sir John Rogerson's Quay and Pearse Street, it was designed by American landscape architect Martha Schwartz, and developed by the Dublin Docklands Development Authority.

The square and the boom-time developments surrounding it weren't long finished when the property bubble burst. With the National Asset Management Agency (Nama) taking control of properties such as Boland's Mill, the Montevetro building on Barrow Street, and the Observatory building on Sir John Rogerson's Quay, the area soon got the name 'Namaland'.

However, while the rest of the country entered recessionary times in the years that followed, a growth spurt would occur in Grand Canal Dock. In October 2008, Facebook confirmed Ireland as a friend, announcing it was going to establish an international headquarters in Dublin. While it initially set up in Fitzwilliam Hall next to the Grand Canal, it would later move to the heart of Silicon Docks, on Hanover Quay. It was soon followed by Engine Yard, Airbnb, Twitter, Squarespace, Riot Games and LinkedIn, among others. By the time Facebook moved to bigger offices at 4 Grand Canal Square in the summer of 2014, the area was employing more than five thousand people in the tech sector.

The Role of VCs

Ever since 1972, when investment firms Kleiner Perkins and Sequoia Capital set up in Menlo Park, venture capital has been the engine driving Silicon Valley forward. In that same year, the Irish government established Foir Teoranta to provide finance for potentially viable businesses that were unable to raise capital from the banks.

Several years later, State-owned venture-capital company Nadcorp was established and, in the five years that followed, it invested approximately £26 million and took shares in fifty-seven companies. It was subsumed by the IDA in 1991.

In the mid-1990s, the Enterprise Ireland Seed and Venture Capital Programme was conceived, on the basis that the private sector on its own would not provide equity capital for high-risk, high-growth companies.[5] In the twenty years since then, the government, through Enterprise Ireland, has committed more than €320 million to seed- and venture-capital funds that have been set up in Ireland.

This, along with the arrival of a number of venture-capital firms (VCs) to the country, has ensured massive growth in Ireland's venture-capital industry. Since 2008, more than €650 million of international funds has been leveraged by Irish-based venture-capital firms into indigenous small and medium-sized enterprise (SMEs). In 2013 alone, Irish firms raised €285 million in venture capital.[6]

The arrival of a number of US VC firms such as Polaris Partners and Silicon Valley Bank boosted the Irish start-up scene further, strengthening Silicon Valley's ties with Ireland. Silicon Valley Bank has invested $50 million into Irish tech and science-based companies and plans to double that. Polaris has also invested tens of millions.

Investor Noel Ruane, a venture partner with Polaris, says Ireland has long been the Internet capital of Europe when it comes to foreign direct investment but, more recently, local start-ups have also started to attract international attention.[7]

The Changing Start-up Scene

In 2011, three Irish students between the ages of eighteen and twenty-one created an augmented-reality start-up called Charm Mobile, on the back of an AIB bank student loan and credit card.

The company's software was a rendering engine, which was licensed by a digital agency working with McDonald's and used in a campaign for the fast-food giant across seven thousand locations in the United States. That three friends, who had not yet graduated from college, could develop a product that would be used to promote one of the world's biggest companies shows how far the Irish start-up ecosystem has come.

One person who noticed the increasing number of start-ups early on was Vincent Lyons. In 2012, he co-founded Dubstarts, a jobs fair for start-ups. As one of the first employees of Startupbootcamp, an accelerator for start-ups, he could see a rise in the number of start-ups popping up, all of which were looking for staff. At the same time, he noticed college graduates were having a harder time securing jobs, with many emigrating.

So Lyons set up the Dubstarts jobs fair with Charm Mobile CEO Enda Crowley, in an attempt to match start-ups with potential employees. Logentries, Intercom, Newswhip, Bullet HQ and Upfront Analytics exhibited at the first fair and a hundred people were signed up to attend. More than three hundred showed up.

The Dubstarts events have since grown dramatically and attract in excess of 3,500 attendees, with up to fifty companies exhibiting. Lyons says it is now 'normal' for people to work in a start-up, whereas, in the past, they wouldn't have even known what a start-up was. He says the idea that a graduate can join a small start-up and help it to develop into a multi-million-euro company is really compelling.

However, Lyons is unsure whether Ireland has what it takes to create the next Facebook or WhatsApp. He says there are lots of people in Silicon Valley with experience in bringing companies to billion-dollar valuations. If you give an Irish start-up €1 million, he believes there is less chance of it being successful than if it were in Silicon Valley, as there is more operational experience in Silicon Valley.

While the growth of start-ups in Dublin has been phenomenal, it seems few can afford to locate in the Silicon Docks area. Grand Canal Dock may be a tech hub, but across the city, in the vicinity of the Guinness brewery, is Dublin's start-up hub. Since 2003, more than 160 technology and digital-media businesses have passed through the Digital Hub. It is from here that start-ups such as US e-commerce site Etsy run their European operations.

A short stroll away is the Guinness Enterprise Centre, which has seen more than 260 businesses pass through its doors since it opened in 2000. One start-up currently in residence is B-smark, which coincidentally shares its birthday with Arthur Guinness. Showing the influence the established tech sector has on the start-up scene, B-smark's Italian founder Nicola Farronato says the reason his start-up is in Ireland is because of Intel's Martin Curley.

Having started out in Dublin doing some voluntary research for the Innovation Value Institute, of which Curley is one of the founders, Farranato decided to set up his own business, which became the first international start-up to participate in the National Digital Research Centre (NDRC) Launchpad accelerator programme. Farranato says he chose Dublin as the location for B-smark, because the start-up ecosystem in Italy and southern Europe is 'effectively non-existent'.

Ireland's Reputation as a Tech Hub

While venture-capital funding has been vital in the development of Dublin as a tech hub, money alone cannot buy the sort of reputation F.ounders and the Web Summit have given the capital city. Founded by Paddy Cosgrave, Daire Hickey, and David Kelly, the two events attract some of the most influential names in the tech world to Dublin each year.

Since 2010, top entrepreneurs have descended on Dublin, including Dropbox founder Drew Houston, Skype founder Nicklas Zennstrom, and PayPal and Tesla co-founder Elon Musk. The conferences help showcase the capital city as a hive of tech and entrepreneurial activity, and it was coming to Ireland for these events that put Dublin on the map for tech firms such as Qualtrics, leading them to ultimately set up offices here.

As well as being home to Europe's largest tech conference, it was also in Dublin that Archipelago, one of Europe's largest communities of entrepreneurs, was born. The brainchild of serial entrepreneur John Egan, the network's entrepreneur events showcased up-and-coming start-ups such as Datahug, CoderDojo and Logentries. These would later go on to achieve global success. Egan says Archipelago was a manifestation of the environment of the time, and it would seem the current environment in some ways is a manifestation of Archipelago, as it gave many an attendee the confidence and inspiration to set up their own business.

Egan, himself a successful entrepreneur, credits Web Summit founder Paddy Cosgrave with legitimising the tech scene in Ireland. He also believes Trinity College Dublin has a lot to answer for in terms of Dublin's development as a tech hub, with people such as Cosgrave and serial entrepreneurs Dylan Collins, Sean Blanchfield and Ronan Perceval all graduating from the university. He says the four entrepreneurs have, in different ways, brought attention to Ireland's tech sector. They also created an energy at Trinity – tech was becoming glamorous at the university.

The Influence of Education

More companies are spun out of Stanford University than any other university in the world. As well as HP and Google, Sun Microsystems, Silicon Graphics, Cisco Systems and Yahoo also emerged out of the Silicon Valley-based university. Instagram was founded by Stanford graduates Kevin Systrom and Mike Krieger, while Snapchat started as part of a class project by Stanford students Evan Spiegel and Bobby Murphy. Other Stanford alumni include the founders or co-founders of PayPal, Netflix, Electronic Arts, Mozilla Firefox, YouTube and LinkedIn.

In Ireland, Trinity College Dublin seems to play this role. The university has produced more spin-outs than any other university in the country, and accounts for one-fifth of all companies spun out of third-level education.[8] It has spun out numerous tech companies, including Iona Technologies, which had the fifth-largest IPO in Nasdaq history when it went public in 1997. Gaming-technology company Havok also emerged out of Trinity, having begun life in the computer-science department there in 1998. Less than a decade later, in 2007, it was acquired by Intel for $110 million. MV Technology started out in Trinity's Vision and Sensor Research Unit. It was sold to Agilent Technologies in 2001 for IR£100 million.

The Spin-out Effect

It is not just universities spinning out companies. The arrival of multinational tech firms in Ireland and the development of an indigenous IT industry has also given rise to start-ups, with employees honing their skills in large companies before going it alone. Ireland's oldest IT services company, System Dynamics, was founded after Tom McGovern left IBM in 1968 and, ever since, tech firms have been churning out entrepreneurs.

Iona Technologies, one of the State's most successful software companies, has spun out more than twenty companies. Annrai O'Toole, a co-founder of Iona, left the company in 2001 to set up software firm Cape Clear. O'Toole has since become the chief technology officer (Europe) of Workday, the Silicon Valley firm which acquired Cape Clear.

Brian Long is another successful Irish entrepreneur who cut his teeth in a multinational – in this case Digital Equipment Corporation – before leaving to set up Parthus Technologies. Niall O hEarcain, a technical director at Parthus, later went on to found semiconductor company Silansys.

History is repeating itself with the new wave of tech firms, with employees from Internet giants such as Google starting to set up their own firms. Gaston Irigoyen came to Ireland in 2009 to work for Google. Now he runs his own business – Guidecentral – a few doors down from the search giant, in Dogpatch Labs. Across the office from Irigoyen sits PulpMaker co-founder Charles Alix, who previously worked for Facebook and Twitter. Boxever was founded in 2012, after three employees left Datalex. Datahug co-founder Connor Murphy previously worked at Sun Microsystems.

Meanwhile, many of the people who successfully founded, developed and sold tech companies are now embarking on new careers as venture capitalists, investing in the next generation of up-and-coming firms.

One such person is angel investor Paddy Holahan. He founded East Coast Software in 1987, and spent almost a decade building up the company. He joined Baltimore Technologies in 1996 and played a pivotal role in growing the business from a $500 million valuation to a public company valued at over $16 billion in the space of four years.

Holahan later founded NewBay, growing it into a leading provider of solutions for mobile operators worldwide, before selling the company for $100 million to BlackBerry maker Research In Motion in 2011. He has since invested in a number of Irish start-ups, including Intercom, Soundwave and WhatClinic.com.

The Impact on Real Estate

The impact of the tech firms in Grand Canal Dock can be seen in Central Statistics Office small-area population statistics from 2012, which were based on the 2011 census. They showed that in one of the buildings in the Gasworks Buildings apartment complex, which is located next to Google's European headquarters on Barrow Street, 94 percent of residents were in employment, the highest percentage in the State at the time. Of those that were employed, nearly all were employers, managers or higher professionals.

The impact can also be seen in the rise in office and residential property prices. Many of the new buildings around Grand Canal Dock had just been completed when the global financial crisis hit. As a result, a lot of them lay empty in the years that followed. However, international investors and companies alike are now swooping in to buy the prime office space, and paying premium prices.

Google has forked out more than €250 million for the four buildings it occupies on Barrow Street, paying almost €100 million for the Montevetro building, which is Dublin's tallest commercial office block. Property pension fund IPUT shelled out €50 million for Facebook's former office building at Hanover Reach, while Hibernia REIT paid €52.5 million for the Observatory building on

Sir John Rogerson's Quay, home to Riot Games and financial-technology firm First Derivatives.

An office-space crunch in the area is starting to put pressure on companies located there, with rents going up and up. A review by commercial property firm HWBC found rates for prime offices in Dublin city centre rose 15 percent in the first half of 2014, with demand for office space 50 percent ahead of the same period in 2013. During the first half of 2014, four out of the top five office deals were completed by US technology companies, including Riot Games, which struck a deal for 4,230 square metres in the Observatory building.

Competition to buy or rent residential properties in the Silicon Docks area is also fierce, and while prices haven't returned to their boom-time peak, they have risen significantly since the property crash. The tech firms in the area employ large numbers of workers from overseas, and most of them want to live close by.

However, a new fast-track planning scheme has been approved by Dublin City Council, which should help address supply problems. The Docklands Strategic Development Zone will give council planners the power to make decisions, meaning projects can be fast-tracked through the planning process. The development of properties in the area under the control of Nama receivers and debtors can now be accelerated as a result.

In the summer of 2014, Nama announced plans to pump up to €3 billion into building up the capital city. The centrepiece of the strategy is a €1.5-billion plan to develop the docklands into an Irish equivalent of Singapore's Marina Bay.

The Future of Silicon Docks

Despite soaring property prices, a lack of office space and a shortage of rental accommodation, the rise of Silicon Docks has been an Irish success story. While firms in the International Financial Services Centre (IFSC) came under huge pressure during the global economic crisis, Silicon Docks flourished, with tech companies making one jobs announcement after another. There is little doubt that the transformation of Grand Canal Dock from industrial wasteland to European tech hub has been one of the great accomplishments of the boom.

However, if Dublin is going to become a truly global technology hub, like Silicon Valley, then there will have to be some changes.

If more start-ups are to emerge out of tech companies, then there needs to be product development, research and engineering taking place. The government moved away from attracting assembly-line-type manufacturing to Ireland, when it saw how easily such firms could up and move. The same could also be said for many tech firms now, which are for the most part carrying out customer-support and sales functions.

There also needs to be more involvement by firms such as Google, Facebook, Twitter and LinkedIn in the Irish start-up community. In this area, things are moving in the right direction. To improve connections between start-ups and global companies, some of the tech giants have started to reach out. For example, Google hosts breakfast briefings every month at The Foundry to help entrepreneurs and start-ups boost their business.

A move towards product development and engineering, along with increased collaboration between multinationals, start-ups, and indigenous industry, will allow the tech community to be greater than the sum of its parts. This will enable Dublin to attract more Silicon Valley giants, and it will also foster the next round of Irish tech titans.

1. The History of the Docklands

Joanna Roberts

On 27 May 1960, a small cargo boat laden with Guinness set off from Grand Canal Dock, destined for Limerick. Skirting Dublin's city centre via the canal's circular ring, she travelled through Portobello and Harold's Cross, before joining the canal mainline at Suir Road and continuing west to meet the River Shannon.[9]

In one respect boat 51M was unremarkable, following, as she did, in the wake of hundreds of other barges used to transport goods between Dublin and the west of Ireland since the eighteenth century. However, this particular voyage had an added poignancy, for it was the last time that the Grand Canal would be used for commercial purposes. First the railways and then the roads had eliminated the need for goods to be transported by water. The cost and time involved in travelling by canal had long outweighed its benefits.

In 1796, when the Grand Canal's circular ring first opened, it had formed the southern boundary of what was then a thriving port city. Every day, ships containing loads such as grain, coal, tobacco, metal ore and sugar arrived at the mouth of the River Liffey, guided on their way by the newly constructed North and South Bull Walls, which had been designed to create a channel into the river.

The Custom House, in use then, had been built in 1791 to replace the original building at Wood Quay, and the new structure had the effect of moving the location of the port away from the city centre towards the east. The quays against which the ships moored and the docks onto which they unloaded their sacks of cargo were also new, having been reclaimed from marshland to facilitate trade.

The thriving industrial economy and the round-the-clock nature of the port meant plenty of work for people who lived in the area. Goods were unloaded into quayside warehouses, with some earmarked for use in the city and others destined to continue their journey west along the Grand Canal. To unload and sort the goods that arrived required manpower, and so the shipping companies and factories employed local men as casual labour. To see a ship come in was to see huge numbers of men head to the quaysides in the hope of being picked to work as a docker for the day.

Over the course of the nineteenth century, industry grew rapidly in the docklands. Coal was the main source of fuel for most households and shiploads of it arrived every day from Britain and further afield. The area was chock-a-block with coal merchants and many people found work digging the coal out of the bowels of the ships.

Coal could also be turned into gas. In the 1820s, the Dublin Gas Company established a gasworks on Sir John Rogerson's Quay, which would produce gas for the city of Dublin for the next 150 years. A large gasometer – a circular metal skeleton surrounding a storage chamber that rose and fell as gas was produced and used – was erected at the corner of Macken Street and Sir John Rogerson's Quay in 1934 and became a Dublin landmark.[10]

The land to the south of the Liffey also provided prime real estate for processing companies. Locating in the docklands meant having easy access to warehouses full of goods and a plentiful supply of water. A sugar refinery was set up to process sugar cane arriving from overseas. A large flourmill was built to serve Boland's Bakery, which was situated on the road between the city and Ringsend. The area also became a centre of glassmaking, thanks to demand for glass bottles to store milk and Guinness, and a good supply of sand and water.

For over a century, the docklands remained a vibrant centre of trade. A new sewerage treatment works at Pigeon House in 1906 meant that the city's sewage was no longer pumped straight into the Liffey. Inhabitants of the docklands were thus relieved of the stench that had blighted the air breathed by their forebears.[11] But the twentieth century also brought other changes, ones that would alter the area beyond recognition.

A Changing Economy

The final delivery of Guinness along the canal marked just one of the ways in which times were changing for the docklands. It wasn't just that goods were being transported away from the docks using different methods; the entire economic ecosystem around the docks was beginning to shift. From the 1950s onwards, technological advancements, increased globalisation and political decisions all combined to fundamentally alter the physical, social and economic fabric of Grand Canal Dock.

First came mechanisation. During World War II, supplies were for the first time transported in large rectangular containers that could be easily stacked, catalogued and transferred from one form of transport to another. The advantages of containerisation over the expense and complexity of employing people to load, unload and store sacks of cargo meant the practice quickly spread. Containers were used to transport goods on ships and then seamlessly transfer them to lorries for their onward journeys, eliminating the need for warehousing and sorting. Suddenly cranes and forklift trucks began to replace casual labourers and a large source of work for dockers ran dry. Between 1975 and 1984 employment in the port fell from 7,403 to 5,200.[12]

The early 1970s was also the time of decasualisation, where the daily uncertainty of whether or not work would be available was replaced with a more regulated system. A dockers' register was set up, with work going only to registered dockers, and other benefits such as a weekly wage instead of daily pay. While this was introduced to give some certainty to employment, the big effect was that many men who had picked up day work for years were no longer required. A large number took payoffs and became unemployed. Yet even for those who continued to work, the prospects were shrinking. When decasualisation was introduced, 550 men were put on the dockers' register. By 1992, there were just forty-two permanent dockers and one hundred in a supplementary pool.[13]

The combined impact of containerisation and decasualisation meant that Dublin, like many other port cities at the time, saw unemployment around the docklands increase dramatically. While there was still work around, opportunities were fewer and further between. People who had for years relied on the docks to provide them with manual work now found themselves unskilled and lacking in employment prospects. Low levels of education and lifetimes of manual labour did little to prepare dockers for other types of work.

Other docklands industries also began to decline. By the 1960s, electricity had become widespread in homes across Ireland, which reduced demand both for gas and the coal required to produce it. What's more, the type of gas used in Ireland was changing. Natural gas, which was viewed as cleaner and cheaper than coal gas, had been found in the North Sea and off the coast of Kinsale, in County Cork. In the late 1970s, the newly created Bord Gáis began production of natural gas from the Kinsale field and, in 1983, a pipeline supplying gas from Cork to Dublin was opened. The infrastructure's completion nailed the coffin closed on the Dublin Gas Company, which transferred its assets to Bord Gáis and shut up shop. More jobs were lost, and the gasworks became obsolete.

Europe also played a role. In 1973, Ireland joined the European Economic Community (EEC), with the promise of free trade, access to funds and reduced dependency on the UK market. However, EEC membership also meant throwing off the policy of protectionism that defined Irish industry. Restrictions and tariffs on imports were now things of the past, and many Irish industries suffered as the economy adjusted to the new environment.

Employment in the docklands declined further as more factories closed down. Others chose to move out of the city and into areas on the outskirts of Dublin, where manufacturing clusters were forming. In the late 1970s, Boland's Biscuits merged with Jacob's Biscuits and moved its factories from Grand Canal Dock out to Tallaght, on the south-west fringes of the city.[14]

All these factors swirled together to form a perfect storm. By the late 1980s, what had once been a noisy, active industrial quarter providing a reliable source of manual work had fallen quiet. As mechanisation increased and factories closed or moved out of town, work dried up. The area fell into rapid decline.

The Docklands Drain

For a community that was so tied to the jobs that came with the area, the industrial decline proved devastating. For years, people lived close to the docks so they could get there quickly and be first in line to be picked for work when the ships came in. Docker 'buttons', which ensured that their wearers were prioritised for any available work, were handed down from father to son, and the livelihoods of entire families were tied to the docks.

However, life was defined by poverty. Work on the docks may have been regular, but it was hard and often dangerous. Food was often scarce. Cyril Deans, who worked as a docker throughout the 1960s, describes his life as a 'coalie' – digging out coal from ships – as 'slavery', with no security, no pension rights and no avenue for complaint if the shipping company decided to dispute your performance. Before decasualisation, wages were often paid – and spent – in the pub. It was an impoverished life: Cyril's father, also a docker, often used to get items out of the pawn shop on Monday, only to have them back in by Friday.

Despite its inherent insecurity, dock work was a family trade that passed down through the generations. When it came to getting picked for work, family credentials were more important than academic achievements. As a result, there was no tradition of valuing high education levels, and school drop-out rates were high. By 1997, only 10 percent of young adults in the area stayed at school long enough to sit their Leaving Certificates and a mere 1 percent went on to third-level education.[15]

When jobs disappear, people follow. As the main sources of employment steadily declined over the second half of the twentieth century, people began to leave: usually those who were young, educated and had a bit of money for a new start. Those who could work moved away, either to another part of Dublin or overseas, leaving behind those that couldn't. The docklands gradually became a neighbourhood of the elderly, poor and unemployed. Social structures disintegrated. What was once a close-knit community where people lived, worked and socialised within tight boundaries became a fragmented echo of its former self. Community ties were broken and the population declined.[16]

In the meantime, social policy in Dublin concentrated on suburbanisation, which aimed to tackle the problem of inner city slums by encouraging people to move out of the city. In the 1960s, the new towns of Tallaght, Coolock and Ballymun were developed in order to provide affordable housing and a better standard of living than that in the centre of Dublin. While all hopes were pinned on the outskirts of the city, those left behind suffered from a lack of investment.

With broken community ties, minimal investment and many of those with ambition leaving the area, little remained for the people in the docklands. By the 1980s, the area was an example of the worst of inner-city living: high unemployment, poor health, low education, high drug use and crime. In 1997, unemployment was endemic, standing at more than double the national average. Drug use and crime boomed, and the docklands gained a reputation as a no-go area.[17]

Urban Dereliction

The economic and social changes of the twentieth century left physical marks on the docklands landscape.

The introduction of roll-on, roll-off freight in the 1950s meant less space was needed for warehousing, and so the Dublin port moved further east, down the Liffey, to its current position. As shipping activity withdrew from the city, it left behind a wave of dereliction along the banks of the river. Because the Liffey cuts straight through the centre of Dublin, environmental and urban degradation along the quays had a stark impact on the city's appearance.

What's more, planning policy at the time discouraged private development of derelict sites. The desire to widen the streets along the quays had an unfortunate side-effect: any application for refurbishment in that area would lead to the site's boundaries being reassessed and its size likely being reduced. Developers were put off, and buildings were left to fall into ruin, with boarded-up doors and broken windows adding to the desolation of the area. Ruari Quinn, who in the mid-1980s was Minister for Labour, remembers the urban blight along the quays as giving the city the appearance of someone with a very bad set of teeth.

Even in its heyday, the area around Grand Canal Dock was far from a green and pleasant land. The geographical boundaries of the Liffey, Grand Canal, the city and the sea meant space was limited, with much-needed housing crammed into narrow roads around Pearse Street and Ringsend, and the Grand Canal Dock area reserved mainly for industry. With the exception of Ringsend Park, which provided an important splash of greenery, open space and recreational amenities were few and far between.[18]

Grand Canal Dock was one of only two industrial areas in Dublin, the second being the area around the Guinness factory at St James's Gate. As the economy declined, factories fell quiet, gates were locked, forecourts abandoned and buildings left to crumble. The closed gasworks on Sir John Rogerson's Quay had left behind a classic brownfield site: abandoned, desolate and neglected. Heavy contamination from years of gas production meant it was very risky for a developer, and no one wanted to touch it.

Unless a ship arrived, the area was quiet. People might visit the old scrap yard at Hammond Lane to make a bit of money but, by the 1980s, the area around Grand Canal Dock was a shadow of its former self. Where once there had been busy factories buzzing with activity, there were now rusty structures standing on land polluted with the by-products of industry that was now a thing of the past. Derelict buildings and patches of wasteland defined the landscape, and much of the area was simply sealed off. A rapid industrial decline combined with a policy of getting people out of the area meant that the docklands quickly became an eyesore.

Housing was also a problem. In the late nineteenth century, cheap housing had been built to accommodate the swelling ranks of workers. Space and budget limitations meant the buildings were simple: one and two-storey terraced houses in narrow streets. Throughout much of the early twentieth century, large families lived in cramped conditions in small houses, with no space to expand. In the 1930s, some areas of the docklands, particularly around Townsend Street, were targeted for tenement clearance. Social housing was built in the 1950s and again in the 1980s. However, a concerted regeneration project was still just a twinkle in a future generation's eye.

Urban Renewal

By the early 1980s, it was clear that Dublin's docklands needed attention. Regeneration projects had been successful in dockland areas elsewhere in the world and provided food for thought for Dublin. There was also a renewed focus on the city of Dublin as it approached its millennial anniversary in 1988. Gradually, planning policy shifted to focus on redeveloping the inner city, including the former industrial areas of the docklands.

The first priority was the north side of the river, around the Custom House. Although then Taoiseach Charles Haughey set aside IR£14 million for new housing north of the Custom House in 1982, redevelopment of the area didn't really get underway until the Urban Renewal Act and the associated Finance Act came into effect in 1986. Both introduced a range of incentives to encourage urban renewal by the private sector.

At the same time, the government, encouraged by businessman Dermot Desmond, decided to actively pursue financial services firms and develop an Irish financial services industry. As a result, it was decided to develop the north quays into a centre of financial services, inspired by the successful regeneration of the Canary Wharf area in London's docklands. The Financial Services Act 1987 provided a number of tax incentives to companies. A 10-percent tax rate was introduced to financial services and was linked to geography: businesses had to physically locate in the docklands to be eligible.

In 1987, the Custom House Docks Development Authority (CHDDA) was set up and tasked with redeveloping eleven hectares of land in the Custom House Docks area. The first building was completed and occupied by AIB in 1990. By the time the authority was dissolved in 1997, 114,000 square metres of office space and 333 new apartments had been constructed.[19]

Although the project was successful, it also had its critics. One factor commonly listed as an oversight was that it had purely economic interests and failed to concentrate adequately on social integration. Little account was taken of local communities, who felt shut out and marginalised by the new businesses in which they could not participate and new housing which they could not afford.

As a test run, there was a lot to be learned from the CHDDA's development of the International Financial Services Centre (IFSC). From an economic point of view, clustering worked. With the right offer, foreign companies could be attracted to Ireland. Moreover, with money and commitment, it was possible to turn a formerly neglected part of the city into a shiny economic hub. But for an area to work as an integrated part of the city, what happens to the people who already live there also matters – the people whose families worked for generations on the docks, and in the gas, coal, glass and chemical industries of the area.

These lessons were to prove crucial when the spotlight of urban regeneration turned south of the river and focused its beam on Grand Canal Dock.

2.
The Redevelopment: The Role of the Dublin Docklands Development Authority

Joanna Roberts

When Seán O'Laoire stands in the middle of Grand Canal Square, particularly on a sunny day, he can barely believe the change that has taken place.

In 1997, O'Laoire authored the first Docklands Master Plan, a template for how the success of the IFSC regeneration could be replicated and expanded to the rest of the docklands area. As both an architect and a specialist in urban planning, his vision was to create a thriving city quarter closely connected to the rest of Dublin. Housing, leisure facilities, and public space would intermingle with office buildings to create a bustling area used by people both day and night.

Grand Canal Dock, with its flagship square overlooking a basin of glittering water, is the physical embodiment of this vision. What was, only fifteen years ago, a poisoned, scrubby wasteland, has been transformed into a vibrant public space surrounded by modern apartments, a five-star hotel, a theatre and an array of bars, cafes and shops.

Waterside seating, fledgling trees, new parks and public art give the area the sense of a gathering place, a picture reinforced by the markets, water sports and festivals that frequently take place.

There are links with history: local landmarks such as Boland's Mill, the Factory on Barrow Street, and the gasworks chimney being prominent reminders of the area's industrial past. Most importantly, however, there are people. Grand Canal Dock is busy once again. Workers from nearby offices, tourists and young families are all breathing much-needed life into the area.

The Origins

Rewind to the mid-1990s. Economically, the IFSC was a success. The Celtic cub was beginning to roar as Ireland's economy started its trajectory of growth. Politically, there was a shift to the left, as the rainbow coalition of Labour, Fine Gael and the Democratic Left were in power.

The Industrial Development Authority (IDA), which had been set up in 1949 to promote Irish exports, had recently refocused its efforts towards the promotion and development of high-quality foreign direct investment into Ireland. Much of this effort was directed towards American companies, which were attracted by Ireland's low corporation tax, EU membership, English language and cultural affinities.

However, Dublin's capacity to house these companies was running low. Ruairi Quinn, who by that time was Minister for Finance, says demand for commercial office space was growing faster than expected. While the traditional model of business parks on the outskirts of the city had worked well for manufacturing companies, he says, it was unsuitable for the new service and knowledge-based industries. Companies were looking for high-quality, modern buildings that were integral to the city, in order to attract a dynamic, urban workforce.

The good news was that what Dublin lacked in existing facilities, it made up for in potential. The size of the undeveloped land in the docklands was equivalent to 10 percent of the area of Dublin that is sandwiched between the Royal Canal on the north side of the city and the Grand Canal on the south side.[20] The problem was that the docklands were so neglected that no private developer was keen to touch them. A bigger solution was needed.

The DDDA

That solution came in the form of a 1997 act of the Oireachtas, which established the Dublin Docklands Development Authority (DDDA). The remit of this semi-state body was to regenerate the remainder of the ailing docklands area not touched by the CHDDA: a 1,300-acre area that included Grand Canal Dock and the North Lotts on the opposide side of the River Liffey.

The DDDA was tasked with securing sustainable social and economic regeneration of the area, improving the physical environment, and continuing the development of the IFSC started by the CHDDA.[21] Crucially, social regeneration was given top billing. This was partly down to the political environment of the time and partly down to recognition that the CHDDA's mandate had been too narrow. Quinn says the DDDA's aim was to achieve genuine urban regeneration, job creation, and wealth creation; it was to be a project based around people rather than just bricks and mortar.

This commitment was baked into the authority's structure. Operating under the remit of the Minister for the Environment, the DDDA comprised a board, a twenty-six-member council and an executive team. Seven representatives from the local community – the community liaison committee – had places on the council, to ensure that people already living and working in the docklands could have a say in every decision taken.[22]

The authority's mission statement was nothing if not ambitious: 'We will develop the Dublin Docklands into a world-class city quarter, a paragon of sustainable inner-city regeneration – one in which the whole community enjoys the highest standards of access to education, employment, housing and social amenity and which delivers a major contribution to the social and economic prosperity of Dublin and the whole of Ireland.'[23]

What lay ahead was a mammoth task. It involved fundamentally remodelling the physical structure and economic fortunes of the area while also raising the living standards of its local population.

Given the track record of the IFSC development, there was little reason for local people to believe that they would get anything out of the regeneration. However, as Sean O'Laoire points out, everything from security to the welcome that newcomers would get depended on the local community. He says the challenge for the authority was to get out of 'do-gooder mode' and really listen to what people had to say.

The authority also had to change the external perception of the docklands. Peter Coyne, who was chief executive of the DDDA from 1998 until 2005, says that for the project to be a success, it had to fundamentally change people's mental map of Dublin by expanding their concept of the city and encouraging them to spend time in an area that was once off limits.

This started with the developers, for whom investing in the docklands had, to date, been a risk too far. For Coyne, the authority's *raison d'être* was to reduce this risk to such an extent that the land became an attractive proposition to private developers. He describes the authority as a temporary interventionist force, stepping in where market forces had failed, in order to kick-start the area's regeneration.

The Plan

The first Docklands Master Plan, written in 1997 and followed by two more in 2003 and 2008, set out a coherent vision for the new docklands. Although the city desperately needed new office space, O'Laoire says it was important to go beyond a short-term vision of an inner-city business park and towards a mixed-use development. The target area was so large that, in his opinion, sustainable development couldn't rely on one function alone. The docklands needed a resident population and visitors who came to the area for entertainment and culture.

A taskforce consisting of professionals ranging from sociologists to engineers came up with a plan to join up the docklands with the existing city, reserving much of the area for housing. The riverside, the other waterfronts and the ground space would be preserved for public use, rather than being sold off to private entities. The grand vision was to create a space with an 'eighteen-hour day' – one that would be used by the public morning, afternoon and evening, and imbue the area with a sense of life and security.

If any indication is needed of the wisdom of these decisions, it is given by the fact that looking back they seem like obvious choices. At the time, they were anything but. Peter Coyne says that the master plan was viewed as experimental, brave and novel. Sean O'Laoire recalls working hard to persuade people of the virtues of a mixed-use development that would achieve a 'perfect synergy' of economic, physical and social redevelopment.

After the vision came the detail. An area action plan and a planning scheme for Grand Canal Dock, covering 38.2 hectares, were approved in 2000. They centred on two key features: the potential of the water bodies and the proximity to the city.[24]

Sixty percent of the area was allocated for residential purposes, with 20 percent of the new housing to be social and affordable housing. This was a first in Irish planning policy, as was the condition that the social housing be mixed in with private housing rather than segregated. This idea was taken from experiences in the Netherlands and Belgium, which showed that integration helps reduce the stigma of social housing.

The planning scheme also divided the land into plots and set out conditions for land use and building dimensions on each one. Importantly, Section 25 of the DDDA Act enabled the DDDA to obtain prior planning permission for the buildings on each plot, which took the pain out of the planning process for developers down the line. As long as developers agreed to the conditions set out in the area plan, they could purchase a plot of land and have their plans approved within two to three weeks by the DDDA. The certainty and speed provided by this process was a key incentive in enticing developers to invest. However, in Grand Canal Dock, there was a major hurdle to clear before any development could begin: decontamination.

Physical Regeneration

In 1998, the DDDA purchased the former gasworks site between Sir John Rogerson's Quay and Hanover Quay from Bord Gáis for €19 million.[25] Years of gas production had rendered the land toxic but, because the contamination was historical, it was difficult to say with any clarity where the liability for clean-up lay. One thing was certain, though: in its current state, no developer would touch it. So the task of decontamination was taken on by the DDDA, whose job it was to encourage development by minimising risk. At a cost of €52 million, the clean-up would prove to be an expensive process for the authority, especially as the estimated likely return was around €40 million. Yet the redevelopment of Grand Canal Dock would not be possible without it.

Between 2002 and 2006, the gasworks site was decontaminated under the supervision of the Environmental Protection Agency. An eight-metre-deep underground wall was built around the perimeter of the affected site, and the land it contained was dug out to a depth of four metres. Some of the toxic earth was treated on site and reused; some was shipped to mainland Europe, where it was cleaned and turned into road fill. The space was then refilled, and the water in and around the site cleaned.[26]

Once that was finished, the DDDA parcelled up the land and sold it on. A rapidly inflating property bubble and rising demand meant that initial estimates on what the site would fetch were quickly smashed. Peter Coyne says that, between cash and property assets, the authority recouped €300 million for the decontaminated land, an unexpected return which could then be ploughed back into social projects.

The DDDA also took responsibility for public infrastructure, which gave additional confidence to developers. They introduced development levies, which raised more than €8 million to help pay for sewers, street lighting and public infrastructure.[27] Decent transport connections were also vital, to connect the existing city to the redeveloped area. In 2001, a new DART station opened on Barrow Street.

In order to open up the water edges for full circulation by the public, the DDDA issued compulsory purchase orders for buildings around the riverside, including U2's recording studio on Hanover Quay. Fans of the band objected heavily to the studio's demolition and bombarded the DDDA with emails and messages condemning the proposal. Eventually, however, it was agreed to build a tower of apartments at the corner of Sir John Rogerson's Quay and Britain Quay, with a new recording studio owned by U2 at the top.

Grand Canal Square

The heart of Grand Canal Dock, and the jewel in the crown of the redevelopment south of the river, is Grand Canal Square, a pedestrianised public space adjoining the dock.

The main function of Grand Canal Square is to act as a focal point for the area. The Bord Gáis Energy Theatre, The Marker Hotel and a variety of bars, restaurants, and shops all combine to create the eighteen-hour day envisaged in the first master plan. Having a space that is used by people outside of core work hours, to host events from festivals to boat races, has generated a natural pull to the area above and beyond commercial activity.

This space, which was designed by Martha Schwartz, features a 'red carpet' of red poles that leads from the theatre to the water and a 'green carpet' of paving with lawns and vegetation that connects the hotel to offices in the area. The 'green carpet' also contains marshy plants in boxes to remind people that the area is reclaimed marshland.[28]

By building Grand Canal Square before anything else in the area, the authority was able to show its commitment to developing the area. It was also able to use the square as a promotional tool to attract development, and it rented an apartment at the top of Charlotte Quay to use as a marketing suite. Peter Coyne remembers looking down at the square, with its lights switched on, and not needing much else to persuade interested parties of the site's potential.

Economic Regeneration

Unlike with the IFSC, companies were offered no special tax incentive to relocate to Grand Canal Dock. The authority had to rely on the quality of the regeneration to attract commercial operations.

The demand for high-quality office space was growing as the economy continued to improve, and the principle of economic clustering suggested that momentum would grow once the first companies decided to relocate to the docklands. However, the authority wasn't involved with end users, just developers.

The ease of planning, the knowledge that investment was being made in the area, and a vision of what the future could hold all served to attract developers to the area around Grand Canal Dock. The DDDA's aim was to generate a supply of commercial and residential space to fill the shortage and help economic growth. Their main intention was to create a critical mass so that the area would take on a life of its own without further public-sector development.

The seeds of a tech cluster came from a different angle. According to O'Laoire, Trinity College Dublin was, around that time, taking the decision to move beyond its walls and create stronger links with the city. In 1999, the university purchased a campus on Pearse Street, now the Trinity Technology and Enterprise Campus. This was set up to provide incubation and development space for technology spin-outs and start-ups.

These days, Grand Canal Dock contains two clusters: tech and legal services. Across the river, the IFSC continues to thrive. More than 40,000 people work – and 20,000 people live – in the docklands as a result of regeneration.

Social Regeneration

But what of the existing community? The authority had promised to create the highest standards of access to education, employment, housing and social amenity – the four pillars of its social regeneration plan. Creating shiny new buildings would only go so far towards achieving this aim.

According to Peter Coyne, the DDDA recognised early on that the old way of life in the docklands was gone forever. The industrial economy had given way to the knowledge economy and the education and training opportunities needed to keep pace. So the authority decided to focus on creating opportunities for the next generation.

It decided to tackle low education levels head on by working in tandem with local schools, at both the primary and secondary levels. Initial attempts to connect with teachers were met with suspicion, until the authority could convince the teachers that they wanted to fund, rather than control, initiatives. Over the course of the three master plans, the DDDA funded everything from psychological tests for young children that led to the introduction of more support staff in schools, to the construction of a brand-new third-level facility, the National College of Ireland in the IFSC.

The results were strong and swift. By 2005, the number of people who dropped out of school in the area, before reaching the age of fifteen, had decreased to 30 percent from 65 percent in 1997. Over the same period, the percentage of pupils that sat the Leaving Certificate increased from 10 percent to 60 percent. In addition, the proportion of people going on to university education increased from 1 percent to 10 percent.[29]

In an area where the indigenous industry had all but disappeared, employment was also a key priority. This was tackled in a number of ways. To ensure that local families could gain in the short-term from the project, the DDDA built a local labour quota into all contracts, which stated that 20 percent of people employed in construction of docklands developments were to be from the local area.

Long-term, the authority concentrated on getting young people into work through apprenticeships and job placements, while also funding community training workshops in skills such as IT, painting and decorating, and sewing. As a result of training and placement schemes, more than 300 young adults from the docklands area now work in the IFSC, up from zero in 1997.[30]

People have also migrated to the docklands. Between 1997 and 2008, the docklands population grew from 17,500 to 22,000, reversing the decline that had been in motion since the beginning of the 1900s. Within Grand Canal Dock, the growth was even sharper: between 1986 and 2011 the population increased from 1,493 to 3,573.[31]

This has been helped by the construction of 11,000 modern homes, 2,200 of which are social or affordable.[32] The decision to reserve 20 percent of housing for social and affordable homes has now been incorporated into national housing policy, a development that Ruairi Quinn cites as one of the major legacies of the docklands regeneration.

The fourth pillar of the DDDA's social regeneration plan – social amenity – was designed to make the area an attractive prospect to new and existing residents. There are new play centres, sports facilities, and health centres. There is waterfront access and new green space at Chimney Park and Hanover Quay Garden. Peter Coyne says that persuading people with children to buy housing in a city location rather than the suburbs was especially important, as it evened out the demographic differences between private and social residents whose homes were integrated in the same block.

The End?

In a very short time, Grand Canal Dock has been transformed from a toxic, derelict wasteland into a vibrant community unrecognisable from fifteen years ago. Thanks to a rising economic tide, a visionary master plan, experience from a previous development, and a strong commitment to community development, the area has a new lease of life and a sense of optimism about the future.

Despite the DDDA's enormous achievement, the authority did not come out of the process with an unblemished record. After several high-profile setbacks, including concern over governance and a high-priced purchase of the former Irish Glass Bottle site, the DDDA was wound up in 2012.

There is also unfinished business. Although much of the development in Grand Canal Dock is complete, some projects, such as the U2 Tower, are yet to see the light of day. What's more, surrounding areas such as the Poolbeg peninsula are still in need of development. In 2014 An Bord Pleanála approved the designation of the docklands as a strategic development zone, keeping its expedited planning status but with control returning to Dublin City Council.[33]

Peter Coyne points out that the DDDA was never meant to be a permanent entity. It existed to create the conditions for development to reach a critical mass and take on a life of its own. What will happen with the as-yet-undeveloped sections of the docklands remains to be seen. While the wind-up of the DDDA can be seen as the end of a chapter, the story of Grand Canal Dock is one that will continue.

3.
Recruiting Companies for the Docks: The Role of IDA Ireland and Enterprise Ireland

Pamela Newenham

When word reached the Industrial Development Authority (IDA) in March of 1989 that Intel was scouting Europe for places to locate two manufacturing plants, the government agency went into red alert. The two plants would bring with them 2,500 new jobs: 1,000 in the assembly of computer systems and 1,500 in the manufacturing of microprocessors.

The IDA believed that winning the systems plant would be good for Ireland, but not transformational. Winning the microprocessor plant, on the other hand, would be transformational.[34] Ireland, however, had no track record of large-scale microprocessor manufacturing. This led to concerns at Intel that Ireland lacked the necessary engineering talent.

Desperate to win the two plants, the IDA commissioned a consultancy group to find expatriate Irish engineers with experience in microprocessor manufacturing. More than 300 engineers were located in a matter of weeks, and 80 percent of them said they were willing to return home for a good career opportunity.[35]

As well as offering a generous package worth IR£87 million, equivalent to 80 percent of the authority's budget that year, the IDA was able to prove to Intel that it would be able to hire the experienced engineers it needed.

The IDA secured both plants for Ireland, and to say it was a good move would be an understatement. In the last four years alone, Intel has spent more than $5 billion upgrading its plant in County Kildare, bringing total investment by the company in Ireland to $12.5 billion. The US computer giant also employs more than 4,500 staff here.

While Intel wasn't the first tech giant the IDA had lured to Ireland, and it would not be the last, it showed the lengths to which the State agency was prepared to go to bring a company here. The evolution of Ireland as a tech hub has largely been due to the work of the IDA and Enterprise Ireland, enticing Silicon Valley giants eastwards, and fostering the development of indigenous firms.

Established as part of the Department of Industry and Commerce in 1949, the Industrial Development Authority was initially created to promote efficiency in the economy. In 1958, with the introduction of the Industrial Development (Encouragement of External Investment) Act, the authority's role was expanded to encourage foreign direct investment and promote exports. It wasn't until 1970, following its incorporation as an autonomous state-sponsored body responsible for all aspects of industrial development, that the IDA became the organisation we know today.

The Focus on Information Technology (IT)

During the 1970s, the IDA identified IT as a key growth sector, and it began seeking out multinational tech firms, encouraging them to set up bases in Ireland. Among the first companies targeted was SAP, which the IDA approached in 1977. It would take twenty years for the authority to convince the enterprise software firm to invest in Ireland – with the opening of its first office in Dublin on 1 April 1997.

Prior to the 1970s, some US companies such as IBM and Digital had located here, but Ireland's tech sector was limited. By the end of the 1980s, Analog Devices, Apple, Dell, Microsoft, Oracle, and Symantec would all have operations here, and in the five years from 1982 to 1987, Ireland saw a doubling of employment in its computer industry.

In 1982, the Irish government commissioned a study of the country's competitiveness in the international marketplace. The study was undertaken by a team from Telesis Consulting, led by Ira Magaziner, who would later become a White House technology advisor during the Clinton administration. The resulting Telesis report criticised the Irish government for spending too much money attracting foreign firms and not enough developing strong Irish companies. The report recommended a reallocation of funding in favour of internationally-trading indigenous industry and Irish manufacturing firms.[36]

The government woke up to the dangers of being too heavily dependent on foreign direct investment in 1990, when the US went into recession. Many of the American tech firms located here were impacted, resulting in the loss of thousands of jobs.

A second study, the 1992 Culliton Report, again urged the government to focus on indigenous industry, and suggested a restructuring of the IDA. As a consequence of the report recommendations, the authority was split into three separate agencies in 1994: Forfás, Forbairt (now called Enterprise Ireland) and IDA Ireland.[37]

The government, through the new agencies, began encouraging the development of Irish businesses, especially software companies. The result was IT firms such as Iona Technologies, Baltimore Technologies, Cúram Software, Parthus Technologies,

Cape Clear Technologies and Trintech, which would later become world leaders in their fields.

Internet Firms Arrive

The IDA also started to redirect its focus away from hardware manufacturing, due to increasing competition from low-wage economies, and instead began targeting the software, data communication, and computer networking sectors.

In 1996, Internet company America Online, more commonly known as AOL, created 500 jobs in Dublin, through a joint venture with German media giant Bertelsmann. It was followed a year later by Netscape, which at the time was locked in a browser war with Microsoft. Netscape, which was the first commercial Internet browser, had been founded just three years earlier, and Ireland would become home to its first software-development centre outside of North America.

While AOL and Netscape formed the second wave of technology firms moving to Ireland, after Intel, Apple, IBM, and Microsoft, a third wave of born-on-the-Internet companies would soon follow. The IDA announced it was going to market Ireland as a key European hub for e-commerce. Bringing Google over would be the agency's first major coup.

In 2003, following lengthy talks with the IDA and several trips to Ireland, management at Google signalled that they had chosen Dublin as the location for the company's European headquarters. Part of the Google delegation that had visited Ireland inspecting sites for offices was Sheryl Sandberg. Recalling the trip, she said the IDA gave her a phone with an extremely short number: 'some number like 3000'. Everyone she met over the next couple of years kept asking her how she got a number like that. Her response: 'The IDA gave me this phone'.[38]

Google's original mandate included a European data centre, multilingual editing, customer-support activities, and financial shared services. However, its Irish operations grew more rapidly than anticipated, to the extent that former Google vice president Nelson Mattos said Dublin could maintain Google worldwide if the lights were to go out in California.[39]

The arrival of Google was a key moment for the IDA, and marked the beginning of its love affair with born-on-the-Internet companies. Also, Google's choice of Grand Canal Dock as a location for its operations would lead to an influx of technology companies in the area.

Befriending Facebook

Shortly after Facebook went global in September 2006, the IDA sent one of its business development executives – John Nugent – to Silicon Valley, to scout out tech firms and encourage them to invest in Ireland. The social network came onto his radar at the beginning of 2007, and he was quickly introduced to Facebook's chief privacy officer, Chris Kelly. Nugent met Kelly at the company's offices in Palo Alto and invited him to a dinner being hosted by the IDA in a Palo Alto hotel. And so began a nearly two-year courtship between the IDA and Facebook, one that would ultimately lead the social networking giant to the Dublin docklands.

Facebook had less than fifty employees and was still a 'scrappy start-up' when Nugent met the firm in 2007. He was unsure how the company would scale up enough to be of benefit to Ireland, and he had bigger fish to fry, namely Disney, which the IDA team was also trying to lure. Nugent also had his eye on other early-stage social networks, such as MySpace, Friendster and Bebo. But he was impressed with the pedigree of executives at Facebook, such as PayPal founder Peter Thiel.

Facebook was looking to internationalise, and Nugent set about learning as much as he could about the company, in order to help it. Very early on, he met the developers who built the network. At the time, they were experiencing issues with people registering with the same name. Nugent became slightly worried when the social network opened offices in London, but refused to give up. His Ireland-based counterpart, Brian Bastible, meanwhile took on the role of showing around the Facebook executives that visited Ireland and helping to introduce them to lawyers, accountants, and real estate agents.

When news broke in April 2008 that Sheryl Sandberg was joining Facebook as chief operating officer, the IDA was thrilled. It knew she was an ally of the organisation, as she had previously approached some of the agency's executives at a conference in California and praised them for their work.

On the day that Sandberg started at Facebook, the IDA arranged for a big bouquet of flowers to be waiting on her desk with a note saying: 'From all your friends in Ireland'.

During the summer of 2008, Nugent visited Facebook's offices nearly every day. At that stage, the company had grown to several thousand employees and had a campus of its own.

In October of that year, with Ireland in crisis following the collapse of its banking system, the good news finally came through. Facebook had chosen Ireland for its international headquarters.

There was a glitch though. The Companies Registration Office (CRO) wouldn't allow the social network to register 'Facebook' as its company name, as another company in the midlands had a similar name. The IDA had to convince the CRO to allow it.

With Facebook up and running in Ireland, Nugent turned his eye to the other social networks. He had a stroke of luck in February 2009, when Twitter co-founder Biz Stone tweeted: 'Explaining to my mom that not everyone on the Internet is trustworthy – even if they are from Ireland'. Nugent sent him a message in response saying, 'I'm Irish and you can trust me. We should meet.'

Stone responded immediately and introduced Nugent to his colleague, Twitter's vice president of business operations, Santosh Jayaram. The following month, Jayaram tweeted: 'The guy who came to visit me from Ireland is helping a few people move furniture at the Twoffices. Very hands on here!' Nugent replied: '@santojay all part of the service from IDA Ireland'. Two years later, the microblogging site announced it was establishing an international office in Dublin.

Innovation Comes Naturally

In September 2009, with the country in deep recession, the IDA decided to go all out to promote Ireland as a leading location for innovation and technology, undertaking its largest-ever advertising campaign in the United States. This campaign, with an estimated budget of €2 million, offered a vision of a tech-business utopia, and featured Facebook's Sheryl Sandberg and Microsoft Ireland's managing director, Paul Rellis.

Previous campaigns to promote outside investment in the Irish economy had focussed on the country's people. Themes had included: 'The young Europeans: Hire them before they hire you', 'People are to Ireland as Champagne is to France' and 'People are to Ireland as oil is to Texas'. The 2009 campaign billed Ireland as a place where 'innovation comes naturally' and was delivered through advertising on CNBC and Bloomberg, in the airports of New York, Chicago, San Francisco, and Washington, and in both the online and offline pages of the *Wall Street Journal* and the *New York Times*. Social media tactics were also deployed to start a conversation about 'innovation Ireland', and the IDA website exhibited success stories from some of its client companies.

Two print advertisements ran in a variety of business publications. The first proclaimed: 'Facebook found a space for people who think in a certain way. It's called Ireland.' The second ad read: 'Google searched the planet for the perfect location for their business. They came up with Ireland.' It was a bold statement intended to counteract negative perceptions about Ireland that had followed the property crash and banking collapse. The aggressive PR campaign received critical acclaim, but the IDA was still under severe pressure.

Trouble Brewing

Up until 2010, the agency's strategy had been built on a value proposition of tax, talent, track record, and technology. The IDA had a number of accolades to bring to the table when meeting with potential companies. According to the 2009 *IMD World Competitiveness Yearbook*, Ireland was ranked first for availability of skilled labour, first for availability of financial skills, and first for corporate taxes.

Ireland's population was comparatively young, with just 14 percent over the age of sixty-five, indicating a larger pool of workers from which to draw. What's more, eight of the top ten ICT firms in the world were already in residence here. Dublin, as the capital city, attracted people from all over the world to visit and live, and could serve as a gateway to talent for the 500-million-strong European Economic Area.

But, on the cusp of the second decade of the twenty-first century, these attributes were not enough. Similar models had been adopted elsewhere, and the cost of doing business was cheaper in those places too. Ireland's competitiveness had declined, and it was quickly becoming clear that the country would need a financial bailout from the EU and IMF.

Two years previously, Barry O'Leary had taken over as chief executive of the IDA and, in the months that followed his arrival, he had to deal with the collapse of Lehman Brothers, the failure of Anglo Irish Bank, the property crash and the financial crisis. The Irish economy was in severe recession and IDA target companies were becoming worried. O'Leary says the agency was getting forty-five minutes to pitch to potential clients, of which they had to dedicate thirty minutes to defending Ireland.

The IDA hit pay dirt in early 2011, though, when Intel, Amgen and Google announced major investments in Ireland. Intel was to begin a $500-million upgrade of its facilities in Leixlip, Amgen announced it was going to purchase Pfizer's manufacturing facility in Dun Laoghaire and Google purchased the two buildings that housed its European HQ on Barrow Street, as well as Dublin's tallest office block, the Montevetro building. The investments showed they were staying in Ireland for the long haul and that the country was still a great place to do business. In June of that year, US gaming company Zynga opened a base in the Dublin docklands, followed by Engine Yard and Twitter in October. The next year, LogMeIn, Hubspot, Ancestry.com, and Indeed came knocking.

Looking for Start-ups

It was in 2010 that the IDA made a deliberate move to attract emerging companies, mainly operating in new media, and offering high growth potential. It began targeting young companies in receipt of venture capital (VC) funding, which were looking to internationalise. The VC funding would give the start-ups room to expand, growing hopefully to the size of companies such as Google or Facebook in years to come. The move was a success, with the IDA successfully wooing start-ups such as Airbnb, New Relic and Dropbox.

Around the same time, Enterprise Ireland also started looking at attracting foreign start-ups, albeit much smaller ones than those the IDA was targeting. On 13 July 2011, the board of Enterprise Ireland approved a €10-million international start-up fund to persuade overseas entrepreneurs to relocate to Ireland and establish their start-ups here. The initiative, watched over by overseas entrepreneurship manager Lorcan O'Sullivan, would support incoming entrepreneurs who had credible, innovative and ambitious start-up plans.

Cloud-based video-surveillance service Camba.tv was among the first foreign start-ups to receive investment from the fund. The company's founder, Marco Herbst, had previously set up online recruitment portal Jobs.ie before selling the site to Communicorp in 2005. Herbst moved to Germany, and it was there he came up with the idea for Camba.tv.

The start-up was originally based in Berlin but, through detailed discussions with Enterprise Ireland, the company relocated to Ireland in September 2011. While Herbst eschewed the Silicon Docks area in favour of Wicklow first, and then Dublin's north inner city, he says it is only a matter of time before the business moves down alongside the likes of Facebook and Google.

While the international start-up fund was established in 2011, Enterprise Ireland's focus on overseas entrepreneurs goes back much further. In 1998, the year that Enterprise Ireland was established by the Industrial Development (Enterprise Ireland) Act, a £1 million Millennium Entrepreneur Fund was established to encourage overseas-based Irish technologists to set up new enterprises back in Ireland. Jointly sponsored by Ericsson Ireland, businessman Neil McCann of Fyffes, Bank of Ireland, and Enterprise Ireland, the fund was a success. By 2000, thirty expatriates had returned to Ireland and started businesses.

Enterprise Ireland Looks West

It was also in 1998, that Enterprise Ireland began looking to Silicon Valley to help Irish firms scale up. While the IDA was busy luring US Internet firms such as AOL and Netscape to Ireland, Enterprise Ireland was seeking out incubator office space in Palo Alto. The city was booming though, and Enterprise Ireland couldn't find any space at an affordable price. It opted to locate 20 miles away, in Campbell.

Over the next four years, about thirty-five Irish tech companies launched products, raised capital and expanded their operations thanks to the Silicon Valley base. One was gaming company Havok. In 2002, Enterprise Ireland eventually got the opportunity to move to downtown Palo Alto.

In 2000, Enterprise Ireland set up its High-Potential Start-Up (HPSU) unit to help high-potential start-ups grow and expand. The HPSU concept stretches back to 1989, when Enterprise Ireland was still part of the IDA. Since 1989, the agency has invested in more than 1,100 companies.

One graduate of Enterprise Ireland's HPSU programme is Gaston Irigoyen. The Argentinian is the founder of Guidecentral, a start-up he set up in Dublin after moving here to work for Google in October 2009. Irigoyen was one of the search company's first five employees in Argentina, and it was at Google that he first became interested in mobile apps. In 2010, shortly after he moved to Ireland, he developed an app for the Fifa World Cup in South

Africa, along with two other people. The app took off, entering the top five in the app charts in numerous countries. As a result, Irigoyen was approached by motor company Kia, one of the sponsors of the World Cup, which paid him to put their branding on the app.

In 2012, with experience in app development under his belt, and noticing the increasing popularity of craft sites such as Etsy, Irigyoen decided to set up his own start-up. The result was Guidecentral, with a mobile application of the same name, for DIY and craft enthusiasts. The app, which allows users to create their own how-to guides, led Irigoyen to a place on the coveted Forbes list of '35 under 35' in 2014. While he moved out of Google, he didn't move far, taking up office space at Dogpatch Labs, on the same street as the search giant, where he has seven full-time staff.

UK technology company Blikbook is another HPSU firm. It relocated its headquarters to Dublin in 2013. The move followed funding of €1 million from Irish investors, including Leaf Investments, Delta Partners, and Enterprise Ireland. Blikbook co-founder Barnaby Voss says the start-up was attracted by the wealth of talent on offer in Dublin, and the presence of a young workforce. The company currently employs ten people and is based on Dublin's Mount Street, on the fringes of Silicon Docks.

Enterprise Ireland and Silicon Docks

A year after Google set up shop in Grand Canal Dock, Mobile Travel Technologies (MTT) was established in Dublin by Gerry Samuels and Paschal Nee. Having supported Gerry's first venture, Gradient Solutions (which was later sold to US firm Sabre), Enterprise Ireland also invested in MTT. The company was initially based at the Digital Hub in Dublin 8 before relocating to Grand Canal Dock in February 2014.

Down the road is another Enterprise Ireland-supported company: Openmind Networks. CEO Alex Duncan says it was investment from Enterprise Ireland that helped set up the mobile-messaging platform and develop it to the point where it has a staff or more than fifty and customers in over thirty countries. At the turn of the millennium, Duncan and his Openmind co-founders Brian Kelly, Pat Flynn and Billy Shekleton, were employed by software firm Logica. With the company in turmoil in 2003, announcing 310 jobs cuts and moving operations to the Netherlands and the Czech Republic, the four aspiring entrepreneurs approached Enterprise Ireland with a break-away idea of their own.

The four had a good story: they were going to offer employment to lots of people left jobless as a result of Logica's upheaval. Enterprise Ireland wanted validation that the business plan was a good idea, so it told the four it would invest if they found someone else to invest too. They did, and ultimately raised €1.8 million from a combination of the state agency and two venture-capital firms. Two months after the money came in, the company received its first order, worth €360,000. As well as providing cash, Enterprise Ireland allowed Openmind to tap into its broad network of offices across the globe, helping to generate overseas business for the start-up.

Funding from Enterprise Ireland also got a spinout from University College Dublin, Logentries, off the ground. The company received a €50,000 investment from the agency's Internet and games fund in 2011. Later that year, it took up residence in Dogpatch Labs, a stone's throw from Google on Barrow Street.

Propelling Ideas into Businesses

In 2010, Enterprise Ireland launched a business-development programme called Propel, aimed at supporting start-ups in the information-technology and life-science sectors.

Participants on the programme received office space and financial support, as well as intensive training in financial management, product and service marketing and international business planning.

It was this programme that led to the creation of Irish software company Boxever. Founder Dave O'Flanagan was working for another well-known software firm, Datalex, when he had the idea in 2010 for a start-up focused on big data for the travel industry.

After securing a place on the Propel programme, O'Flanagan set about developing his idea for Boxever. Towards the end of 2011, O'Flanagan realised he couldn't do everything in the business by himself, and sought out his former Datalex colleagues Alan Giles and Dermot O'Connor to join him. They did, and the company was able to participate in the NDRC's Launchpad programme as a result.

O'Flanagan says Enterprise Ireland's help didn't stop at the end of the Propel programme. The agency was also part of the start-up's first funding round, along with Delta Partners and Bloom Equity in 2012. This money enabled the team to rapidly expand from three to twelve, outgrowing its office space in Dogpatch Labs at the heart of Dublin's Silicon Docks. Having raised $6 million in a funding round led by Polaris Partners, the tech firm now occupies a large office on Tara Street, a short stroll from Silicon Docks. It also has an office in Boston.

Becoming Europe's Tech Hub

There is no doubt that both the IDA and Enterprise Ireland have played huge roles in getting Ireland's tech sector off the ground. Not only did the agencies do the heavy lifting in convincing businesses to locate in Ireland, they also helped businesses get up and running, and fostered their growth.

While the IDA was attracting tech giants and helping them to expand, Enterprise Ireland was encouraging the development of indigenous firms and start-ups.

With their combined efforts, there is now a strong multinational tech presence in Ireland, with ten of the top born-on-the-Internet companies located here, and a scaling indigenous tech sector too.

Many of the firms – from Google to Logentries, Facebook to Openmind Networks, Twitter to Mobile Travel Technologies – are located in Silicon Docks. The IDA and Enterprise Ireland have placed the area firmly on the map as a European technology hub.

4.
Game-changer: Google Moves to the Docks

J.J. Worrall

One of the many elements of David Denby's job is spotting plot holes. Narrative problems are exposed and ill-conceived character decisions highlighted to his readers. This is perhaps one of the main reasons the *New Yorker* film critic can barely comprehend the logic he employed within his dot-com bubble diary, *American Sucker*.

Denby, a sometimes investor, wanted to make $1 million on the stock market to buy his soon-to-be ex-wife out of their Manhattan condo in 1999. To do so, he invested heavily in tech shares. He lost out. Badly. Denby, though, was far from the only sucker in town. The end of the 1990s and the start of the next decade was a time littered with tales of investments gone wrong.

Another man whose name is almost always thrown into the mix when discussion turns to dot-com-era mistakes is George Bell, the former chief executive of the online news source and search engine Excite. In 1999, he decided to decline an offer to buy a company by the name of Google for $750,000. Moreover, this was after the asking price had been brought down from $1 million.[40]

Google, as a large portion of the globe knows by now, was founded by Stanford students Larry Page and Sergey Brin in 1998. It was their second stab at creating a search engine after another – which went by the moniker BackRub – suffered bandwidth issues.

Andy Bechtolsheim, the co-founder of Sun Microsystems (a company which played a bizarrely large part in getting Google to Ireland), was their first investor, writing Google a cheque for $100,000 before the company was even incorporated.

At the time, very few people seemed sure of which web companies had potential for growth and which were built on sand. A number of European countries decided to withdraw foreign development offices from Silicon Valley, wary of getting involved with another dot-com-type blowout.

The IDA stayed though, with the agency's director of operations in California, Dermot Tuohy, insisting there were still plenty of potential partners for Ireland in the area.

He kept knocking on the doors of PayPal, eBay, Overture (which would later become part of Yahoo!) and, of course, Google. By mid-2002 Tuohy – described, in a complimentary sense, by one person who met him at the time, as a 'grizzled veteran'– had taken several meetings with the company, which now has become a verb for Web searches, never mind one of the biggest names in Silicon Valley.

The IDA's Gus Jones also worked on the Google 'account' in California at the time, while, back in Dublin, with the development agency now convinced it wanted to get the company to the Irish capital, John Bolton acted as project manager for the operation.

The Google they were trying to woo at the time had 500 employees and relatively modest (compared to now) revenue of $40 million per annum. The efforts of the three men, alongside recommendations from senior Google staff members who had previously done business in Ireland, convinced the company to take Dublin seriously as a potential location for its European headquarters.

In October 2002, Google sent to Dublin a trio consisting of chief financial officer George Reyes, corporate controller Pietro

Dova, and Ian Cunningham, a consultant the company employed for its site-selection committee. The itinerary that followed was, according to the IDA's Jones, 'typical' of the kind that had attracted previous technology companies to Ireland.

They went to visit the Dublin bases of companies like HP, Symantec, SAP, Oracle, and Citibank, as well as a number of data centres in south Dublin, which, for the most part, had been mothballed in the wake of the dot-com crash.

The visit focused mainly on suburban office parks, as, up until this point, all major technology brands which had set up shop in Ireland had done so on the outskirts of cities. Bolton remembers that after that visit, Reyes was positive about the idea of coming to Ireland.

Both Reyes and the company's chief executive, Eric Schmidt, were among those senior Google figures with Irish business experience from their time with Sun Microsystems, which had an engineering base in Dublin.

Switzerland Beckons

However, this was a two-city race for Google and for Dova, the company's corporate controller, the preferred destination was the Canton de Neuchâtel, in Switzerland. Contact continued between Google and the IDA over the coming weeks but, eventually, the news filtered though that the company had chosen to go with Switzerland. In addition to Dova's support for the decision, several of Google's backers had Eastern European heritage and liked the idea of the company's European headquarters being closer to that region. That's what they told the IDA anyway.

Nonetheless, the agency's representatives in Dublin and California decided they weren't going to accept the decision. They continued to press Ireland's case on several fronts, with those mothballed data centres and Ireland's favourable corporate tax regime front and centre in their arguments.

At the time, a data centre to meet Google's needs would have cost in the region of €50 million to build anywhere in Europe. Those gathering dust in Dublin were available for somewhere in the region of €5 million. Considering Google's relatively limited financial muscle at that time, it was a facet of the Swiss decision that didn't make sense to some within the company, as there were apparently no available data centre facilities in Neuchâtel.

Google thought about it, but still decided that the best they could do was promise that, when they were setting up their next site in Europe, Dublin would be at the top of the list. Again, it was a viewpoint the IDA couldn't agree with. Previous experience with companies which settled in Ireland such as IBM, Microsoft and Intel indicated that once a company set up in a country, it expanded there. They doubted that, if Google ploughed funds into an operation in Neuchâtel, it would choose to start afresh in Ireland a few years down the line, rather than building on its investment in Switzerland.

Jones continued with requests to meet some of Google's top executives and, eventually, he got to sit down with some of the decision-makers in the company's Mountain View offices in California. He put Ireland's 'clear, upfront' tax regime on the table as a positive against possibly negotiating rates in Switzerland. He introduced them to a US company which had offices in Ireland that were thriving, while a larger Swiss arm of the company struggled to justify its existence. With enough doubts planted in the minds of the people opposite him in the boardroom, Jones would soon get confirmation that Google was to revisit the decision.

On 8 January 2003, an Austrian by the name of Gerald Aigner

arrived to inspect those near-vacant data centres in detail. Aigner's employers required persistent questioning and almost clinical examination of the facilities, and one data-centre manager apparently communicated to the IDA complaints about the Austrian's behaviour. But dealing with Aigner, described by *I'm Feeling Lucky: The Confessions of Google Employee Number 59* author Douglas Edwards as Google's 'flaming sword of frugality', was more than worth it. By the time he left Ireland on 10 January, the company had all but decided it should take space in three of the available data centres.

Then, on 19 January, some of Google's top brass arrived in town. The visiting party included Reyes, Cunningham, George Salah, who was vice president for facilities, board member David Drummond, vice president of global online sales and operations, Sheryl Sandberg and Adam Freed, who headed up the company's international product-management team.

Jones picked up Sandberg and Freed from Dublin Airport, as the pair had arrived on a later flight than the rest of the group. He finds it funny to look back now and see the influence the pair would have in bringing more employment to Dublin years later in their subsequent roles with Facebook and Etsy, respectively.

The group from Google looked at business parks in Cherrywood, Parkwest and Citywest in the south of Dublin, as well as Eastpoint and Blanchardstown in the north. Back in the city, they viewed the Digital Hub, a modern workspace close to the city centre which now houses 900 people and is the location for the European headquarters of companies such as Eventbrite and Etsy.

Google's property advisors at the time also alerted them to an alternative location, identifying the potential of a number of buildings on Barrow Street owned by developer Liam Carroll. Walking distance from the city centre, the location was seen by the company as having the right mix of factors to attract the type of employee they wanted in Dublin. In their California offices, Brin and Page had encouraged a college campus-style atmosphere, the likes of which was alien to Irish offices at a time when a foosball table in the canteen was about as leftfield as companies were willing to go.

The visitors decided that, once the building – which was still under construction – was complete, they would rent 60,000 square feet of Gordon House on Barrow Street. It's a choice that is still seen by those in the IDA as a seismic shift for investment in Dublin. One which they, and many others, including senior Google employees, feel was directly responsible for many other Silicon Valley names, such as Twitter and Facebook, choosing to set up shop nearby.

To bookend this particular scouting mission, the president of Dublin City University, Ferdinand von Prondzynski, hosted a dinner on 21 January 2003 for the visitors, and the decision was virtually confirmed that night to IDA officials. Nine days later, the *Irish Times* reported that the delegation's visit indicated a deal was close for the company to set up a European sales, marketing and technical-support centre in Dublin.[41]

This wouldn't be confirmed for another six weeks, though, with estimates suggesting that the company could employ just over 200 people once it moved into Barrow Street. Now the hard work began.[42]

In April 2003, another 'landing party' came to the Irish capital to begin to build a team. It consisted of Freed and Bryan Schreier, who was at the time a senior director for international online sales and operations. The pair sat in a rented office on Harcourt Road, interviewing potential candidates over the course of three days, from 8 AM to 6 PM. Online and newspaper adverts for the roles only mentioned a 'leading online search engine'. But, in the aftermath of the headlines that followed Google's decision to choose Ireland over Switzerland, it was clear to everyone who walked through the door what company they'd be working for.

Google's First Employees in Ireland

Oisin O'Mír, Aoife Wynne, Caroline Dowling, Matt Doris and Janice Azeb were among the dozens upon dozens of candidates to sit in front of Freed and Schreier. Each was called back a second time. Come the following Monday morning, they would be the five people that made up Google's Irish office.

For anyone who has taken a stroll through the company's current docklands base, viewing the facilities that greet the 2,500 or so Googlers (yes, the term is used in the office alongside 'Googliness' and other variants), it's hard to imagine the initially tiny remit of the company's operations here.

Today, as you go from floor to floor, you'll pass a restaurant serving food from around the world, you'll see all manner of games – from chess to pool – being played, and you'll notice a gym, a swimming pool, gaming consoles, toy cars rumbling along the floor, wonderful views of the city and a glass skybridge that connects three of the company's four buildings in the area.

The original offices on Adelaide Road were at least dotted with the odd lava lamp and some artwork on the walls. The five recruits were watched over by Christie Cooley, a trainer with the company who had been brought over from California. Their original day-to-day work involved deciding whether to approve ad after ad after ad.

They examined keywords and reviewed whether each possible advertiser's website stood up to existing policies for Google

AdWords. A repetitive, basic pattern to their workdays began to emerge. That relative quiet and moderate monotony would quickly become a thing of the past with the commencement of a hiring blitz.

What was once considered an incredibly spacious office for five people saw wave after wave of new recruits start to pour in on a monthly basis as the promise of larger digs on Barrow Street grew closer. Alongside reviewing ads, workers were soon asked to help out on facilities work, unpack and assemble desks, chairs and cabinets, get servers up and running, make supermarket runs for sandwiches, and even head to DID Electrical to buy a fridge.

Those waves of hiring were based around getting multitaskers, people with the ability to adapt. Eight to ten interviews were often required for senior employees.

When recruiting German, Dutch or Spanish employees, Google went to Berlin, Amsterdam and Madrid, speaking with recruiters there, finding the best people each city had to offer, and bringing them back to Dublin.

Regular visits from senior executives saw positive reports going back to head office in California. As the 'insanely quick' – as one employee puts it – hiring drive and expansion of the Irish office's remit continued, the original offices were soon packed, with almost every desk shared by four workers. Each conference room was filled up by a dozen or so more individuals, and even some floor-space was occupied by employees huddled over laptops, warning those walking by of stray extension leads, and hoping no one stepped on their lunches.

The company took space in a larger but 'somewhat dingy' office at Seagrave House on Earlsfort Terrace. Numbers swelled and, while there was an appetite to foster the collegiate atmosphere the company's founders traded on, in these early days there also was an ambition to do something other than simply recreating the atmosphere of the California offices minus the sunshine.

The flow of employees from different cultures around Europe created what those in the company at the time say was a unique atmosphere compared to other multinationals dotted around the country. With a huge number of new hires arriving in Ireland for the first time, the social element of the company became hugely important.

The fridge housed bottles of beer each Friday for staff to sip as they talked over the week before. Sandberg, described by colleagues at the time as a 'very clear thinker' and a remarkable delegator with a tremendous ability to listen, was a regular visitor to Dublin. Indeed, she once livened up proceedings at Thank God It's Friday (TGIF) drinks by gathering everyone in the company around to meet a visitor to the offices. Out walked U2 frontman Bono. Whatever feelings some employees may have had towards Bono at the time, it still impressed most of those gathered that Sandberg could call on him to give a speech over Friday evening beers.

By March 2004, the company's California offices had moved to larger digs at 1600 Amphitheatre Parkway in Mountain View, a building christened the Googleplex. August 2004 saw the Google unveil its initial public offering.

That winter, progress marched on in Dublin too, as employees finally moved into the second floor of Gordon House on Barrow Street. They walked through the doors on the first morning there to be greeted by a building that, outside of the space they would occupy, was said to be a 'complete construction site'.

They saw construction workers plastering walls and painting ceilings. Everywhere except the second floor was a hard-hat zone. There was a lack of light in some corners, and one area of the office was called 'The Pit' due to its murkiness in the first few days.

As each new set of ten, twelve or twenty desks was opened up, there was a stampede to claim them.

On 7 October, the company formally opened its European HQ on Barrow Street, with Tánaiste Mary Harney welcoming Brin, Page and other senior Google executives to the country. Workers were said to be buzzing around the offices on motorised scooters. Brin said his original thoughts about coming to Ireland were 'that we would have easy access to Guinness and it was the closest stopover en route to Europe from California'.[43] At that point, the company employed more than 150 people, with numbers increasing weekly.

In January of 2005, a global company conference in San Francisco saw the vast majority of the Dublin office decamp to California, where they hit the ski slopes in the north of the state, near Lake Tahoe.

Herlihy Becomes Head

The company's sheer size was emphasised by the fact that the trip meant they booked up the whole resort for three days. It was on the slopes there that many of the company's employees first had a chance to speak with John Herlihy, the new chief of the Dublin offices.

Having lived in California for twelve years and watched the tech industry bounce back from the dot-com meltdown, Herlihy said it wasn't until 2004 – as Web companies actually started to make money as opposed to just being intriguing opportunities – that he decided it was an area he wanted to involve himself in.

Herlihy had worked at companies including Adobe and Oracle during his time in the US, but the Limerick native felt it was time to head home. In late 2004, he met Sandberg and other Google executives and began to discuss the idea of heading up operations in Dublin, taking on the role of director of online sales and operations for Europe, the Middle East and Africa (He is now the vice president of Google's global advertising operations). There followed what reports have described as a gruelling, twenty-part interview process to confirm he was the man for the job.

By the time he was in place and skiing in Tahoe, Herlihy was certain that the offices he had taken over had a lot more to give. In conversation, Herlihy remembers a company that at that point was hiring talent capable of much more than what was being asked of them. He began to talk to Sandberg about the possibility of increasing employee numbers and the remit of the Dublin offices. It was a visit from the executive chairman, Eric Schmidt, to a rainy Dublin in late January 2005 that proved pivotal.

Over dinner on the 27 January, Herlihy talked to Schmidt about the possibilities that Dublin presented in terms of sales and technical expertise, emphasising the increasing revenue that was being created from the European HQ. Herlihy said the opportunity to attract the right talent to Dublin was greater than they had imagined. With a young, talented, multilingual staff ready to expand into new markets, it was time to take a gamble.

Herlihy wasn't sure how Schmidt would take to the idea. Having seen the company grow from the initial five people checking ads to where it was on that January day, Schmidt told Herlihy to go back to the owners of Gordon House, take the whole building, increase the remit of the Dublin offices and hire as needed.

Suddenly, as Herlihy tells it, they went from a situation of having one 'half-kitted-out' floor of a building to being able to create a long-term plan. Instead of his collection of young, capable managers being asked to take responsibility for sixteen or seventeen different things at once, they could concentrate on a few key functions as the overall management team scaled up drastically within months.

Despite Schmidt's decree that all of Gordon House could be rented out, there was discussion about moving some, or all, of the company's functions to an alternative site outside the city centre. Eastpoint Business Park, Blanchardstown and Citywest were discussed, with Herlihy saying there was a 'definite' opinion formed by some within the company that they were best-served to move to a more easily scalable premises away from the docklands. Herlihy spent quite a large amount of his time convincing those dissenting voices it would be the wrong thing to do. He argued that a young workforce, full of people quite often new to the city wanted to live and work there.

The decision was soon made that Barrow Street offered enough room to expand. That December, Herlihy announced that the Dublin operation was increasing its employee base by a massive 700 people, pushing the number above a thousand for the first time, as it looked for 'graduates with technology or languages qualifications'.[44]

Amid great fanfare the following November, it was announced that, over the coming twelve to eighteen months, another 500 sales, customer-support, IT and management jobs were to be created. However, there seems to be some overlap with the previous announcement, as estimates said this hiring push would bring the number of employees on Barrow Street to approximately 1,300.[45]

Those who worked in the company at the time don't remember anything resembling the thriving tech community that now runs through the docklands. They do remember having to get used to Google's policy of 'deliberately enforcing change', as one employee puts it. Shaking up departments and responsibilities to keep people on their toes, just when they had become comfortable within their jobs, was a regular occurrence.

Employees were encouraged to raise any ideas, concerns or

queries with senior managers. Even the most senior. Top management in Dublin were copied in on emails from Irish employees to Page and Brin, a development which sometimes elicited initial panic, but which almost always saw a positive response forthcoming.

The company continued to expand its workforce in Dublin throughout Ireland's latest recession, and this era saw Google's offices in the docklands expand to include not only Gordon House but also Gasworks House, for a combined annual rent of about €8 million. In 2011, the company made a decision to purchase its entire Barrow Street campus as well as two other buildings in the area. For a combined €226 million, the deal included the Montevetro skyscraper – Dublin's tallest commercial building – as well as the Grand Mill Quay building. An impressive glass bridge linking the Montevetro building with Gordon House and Gasworks House is now one of the most recognisable features of the docklands.

By autumn 2012, the company's Dublin workforce had reached over 2,500 people. Early the following year, the company helped boost the infrastructure behind Gmail, Google Maps and the search engine itself with a €75 million investment in a new data centre in Profile Park, Clondalkin, employing thirty people.[46] The company continued its investment in Ireland that year with the creation of The Foundry, a 15,000-square-foot digital innovation centre costing €5.5 million.[47]

The Issue of Tax

For every investment the company makes and every euro it ploughs into its Irish operations, it's likely that the first public response will be to ask just how much tax Google is paying. With Ireland's corporate tax rate of 12.5 percent – less than half the rate in some other European nations – an established source of ire from other European countries as well as the US, there's no doubt it has been a large part of the conversation for companies arriving into the docklands and other areas.

Google, Facebook and Apple, in particular, have come under scrutiny from a variety of world leaders and politicians over their decisions to base much of their tax affairs in Ireland. That Google is one of the most profitable, visible and established American companies in Ireland availing of the tax regime here does ensure plenty of questions about it whenever Herlihy is interviewed, or any time one of Google's senior executives arrives into Dublin.

The machinations of Google's tax dealings in Ireland are undoubtedly complex. In November 2005, the company said it had 'significantly lowered' its global tax bill for the first three quarters of that year by about €100 million, thanks to its Irish operation. This was at a point when turnover for the Irish operations stood at €603 million for that period. An after-tax profit of just over €2.7 million was announced, with the company reported to have paid corporation tax of €1.6 million.[48]

In September 2013, it was reported that Google Ireland's turnover for 2012 was €15.5 billion, and that it had made pre-tax profits of €137 million. The Irish arm of the company paid €17 million in corporation tax in 2012, leaving after-tax profits of €120.2 million.[49]

Google's most recent accounts, filed at the Companies Registration Office in July of 2014, show the company's revenue climbed 10 percent in 2013 to €17 billion, with pre-tax profits of €189.1 million. Google paid €27.7 million in Irish corporation tax during 2013, and after-tax profits were €154.5 million.

Google's Irish unit reported 'administrative expenses' – royalties paid to other Google entities – of €11.7 billion for 2013, up from €10.9 billion in 2012.[50]

Google Ireland won't say that tax isn't a deciding factor and that it doesn't come up in conversations whenever they're thinking about expanding here. What they definitely will say is that, as a business they have to make a profit, and low taxes only help in that endeavour.

Access to talent, though, is the mantra for any company member asked about the tax question. Ireland, of course, has seen an all-too-steady flow of young, talented workers filing up in airport queues to move to Australia, New Zealand, Canada and beyond during the aftermath of the 2008 bank bailout and the recession that followed. However, the company never struggled to find the right talent, with an estimated two-thirds of the Dublin employee base coming from outside Ireland.

Taxes are low in Ireland for a company like Google, but they argue that's a situation that could change, and they would still be likely to remain in Dublin, in an environment that gives them access to several markets, a time zone that suits the idea of a 24/7 business (since Ireland is, for the majority of the year, eight hours ahead of Mountain View), and an established batch of employees.

The government set the rules, they'll argue, and state that Google will adhere to laws in any country in which it operates. Ask the main people at the Irish offices of Apple, EMC, Intel, LinkedIn and Facebook the same question and you'll likely get the same answer. One Google employee said it's akin to the current motorway speed limit here of 120 kilometres per hour: if that's what you're allowed to do, then that's what you will do.

Google seems to have slowed down after its early bursts of hiring, and the company has sat with an employee base of between 2,500 and 3,000 for some time now. Will it expand? There's certainly room to do so on Barrow Street. The company says that where it can acquire more customers, it can create more demand for work.

As the European economy and, in particular, Ireland, ekes its way slowly out of recession and further away from the latest economic downturn, positive business forecasts should hopefully equate to more money moving around out there for Google to earn and, in turn, create opportunities in Dublin.

In the Dublin office now, more than sixty-five languages are spoken by employees from sixty different countries. They're having to speak about all types of Google innovations too. These days, the company is involved in areas as varied as modular mobile phones and intelligent home heating, to wearable devices and gesture recognition.

Although its tax position raises eyebrows in many quarters, the company has been a hugely positive presence in Ireland and, in particular, the docklands. Aside from the time Google wiped Herlihy's home county of Limerick off the map, things have actually been relatively controversy free.[51]

Well, they didn't actually annihilate an entire county, to be fair. They were simply accused by a local politician, Fianna Fáil TD Niall Collins, of denying Limerick's existence on Google Maps. It was, however, just a slight technical hitch with the tool's zoom function. Google's expansion plans for Ireland aren't quite that drastic.

5.
Social Media's Footprint

Elaine Burke

As the Eurozone went into crisis in early 2009, and Ireland's economy dipped into decline, the first trickles of an uplifting wave of investment from the international tech sector were already spilling into Dublin's docklands.

Though the landscape was still scarred with visible reminders of the property crash – such as the empty shell of an unfinished eight-storey property on North Wall Quay – across the river, Grand Canal Dock was going from strength to strength. Google, currently one of the country's biggest employers, had been enjoying a docklands location since 2004, but it was the arrival of another born-on-the-Internet company in 2008 that sparked an Internet influx in the area.

Friend Request Accepted

Before 1 billion monthly active users. Before a plate of food had ever been Instagrammed. Before there was even a 'Like' button. Before all of this, Facebook came to Dublin. On 2 October 2008, Facebook announced it was setting up in Ireland. Dublin was chosen to be the site of the social network's international headquarters at a time when it had a mere 100 million active users.

The Irish office initially created seventy jobs for Dublin through an investment supported by IDA Ireland. An office in

Hyderabad, India, establishing a footprint in Asia, would follow. As would millions of users, billion-dollar acquisitions and an initial public offering (IPO).

At this point, Facebook was already the world's leading social network, having overtaken MySpace as of April 2008. Membership in Ireland grew in line with global trends as the social network attracted a more mature twenty-five-and-over crowd moving on from Bebo.

As its membership was increasing, so, too, were its staff numbers in Dublin. Facebook quickly grew out of its first digs in Fitzwilliam Hall and took up residence in Hanover Quay with more than seventy employees. This was now the official site of Facebook's EMEA headquarters and was opened by Tánaiste and Minister for Enterprise, Trade and Employment Mary Coughlan in October 2009. By this stage, Facebook was playing host to more than 300 million users worldwide, including over 1 million in Ireland.

At the time, it was expected that Facebook would double the size of its Dublin operation by the following year. True to its word, the company had reached 200 employees by December 2010, when it announced it would begin recruiting for a further 100 jobs.

Life inside Facebook

Inside the Dublin office, the serious business of powering the world's biggest social network was mixed with the serious fun of socialising with colleagues. Instead of Casual Friday, they had Corporate Friday, where employees wore suits to work. Meals were catered and the kitchens were stocked with unlimited snacks. There were regular barbecue parties on the rooftop overlooking Grand Canal Dock and bands playing sold-out gigs in the city would be invited to play exclusive sets for staff.

The online network for friends encouraged its employees to refer their mates for jobs and, by all accounts on Facebook Dublin's own Facebook page, there was plenty of camaraderie to go around. They celebrated with Refer a Friend Friday, doling out cookies and framed photos of employees with friends whom they had successfully recommended.

Before Facebook introduced the Timeline to users' profiles, updates were posted to a Wall. Facebook Dublin had its own physical Wall, commanding visitors to 'Write Something . . .' (under which someone had clarified, in true Irish style, 'DEADLY'). The most famous scrawl on the Facebook Dublin Wall came from the company's founder: 'First time in Dublin. May 27, 2011. Good to meet all you folks! Mark.'

The capital received a number of high-profile visitors in May 2011. There was the first-ever visit by a British monarch to the Republic of Ireland, when Queen Elizabeth II and the Duke of Edinburgh stopped by, and a visit from US president Barack Obama.

After Obama and the Queen had shut down large sections of the city, Mark Zuckerberg was spotted in his trademark combination of hoodie, jeans and trainers, strolling from his temporary abode in the Dylan Hotel on Baggot Street to St Stephen's Green, chatting on his iPhone most of the way, with a security detail following close behind.[52] Zuckerberg was fresh from eG8, a tech forum preceding the G8 summit in Deauville, France, where he briefed world leaders such as Obama, German chancellor Angela Merkel, French president Nicolas Sarkozy and UK prime minister David Cameron on Internet issues, wearing a shirt and tie for the occasion.

On arrival at Facebook Ireland, he was presented with a hurley, and the team had arranged for Dublin band Bell X1 to play a set on the roof of the EMEA HQ. Naturally, this private gig was broadcast on the Facebook Dublin page. Bell X1 front man Paul Noonan described the office as full of beanbags, massage chairs and free M&Ms.[53]

Of course, Zuckerberg also found time for a pint or two on his visit to the capital, and it was even rumoured he held a secret meeting with the Web Summit crew at the Lost Society bar on South William Street.[54] One Irish punter met the famous founder enjoying a pint of Smithwick's at the trendy Village bar on Wexford Street and took a photograph to prove it. With a dash of trademark Irish cheek, he chose to share this snap on Facebook's rival – Twitter.

At the time, twenty-seven-year-old Zuckerberg was said to be worth $15 billion and his company was valued at $80 billion ahead of an expected IPO the following year.

Facebook Continues to Expand

Facebook reportedly began considering an expansion in Dublin as early as January 2012. Among the options being considered were the former Bank of Ireland headquarters on Baggot Street and two office blocks in the south docklands close to its existing location.[55]

The social network became a publicly traded company following its IPO in May 2012 and, later that year, it reached a milestone 1 billion active users. Staff back in Ireland reaped an €11.84-million windfall, with share-based payments working out at an average of just under €30,000 for each of the 400 staff members.[56]

Revenue at Facebook Ireland increased by 70 percent in 2012, to €1.789 billion – 47.7 percent of the social network's global revenues. The growth kept coming and another office move was needed in February 2013, when Facebook Dublin finally outgrew its Hanover Quay location. Another hundred jobs were created and recruitment began – as did the search for suitable office space to hold all of these employees.

The IDA worked with the National Asset Management Agency (Nama), the Irish State agency charged with obtaining the best financial return for the State on its land portfolio, to find a suitable property. Nama advanced funding to complete the site at Grand Canal Square and facilitate the transaction. The agency worked closely with Facebook and the IDA to meet the fast-growing company's requirements and, in November 2013, Facebook's move to 4 Grand Canal Square was officially announced.

Facebook's latest Dublin HQ is a 120,000-square-foot building with the capacity to accommodate up to a thousand workers – doubling the size of its operation and making it the largest of any Facebook office outside of its headquarters in Menlo Park, California. The firm enlisted legendary architect Frank Gehry to design the new office interior, having already deployed his iconic style for its expanded Silicon Valley campus.

It was the city's biggest property deal of 2013, according to commercial estate agents HWBC. In May 2014, property-investment firm IPUT swooped in to snap up the soon-to-be-vacated Hanover Reach office for €50 million.

The Faces behind Facebook Ireland

Like the social network it helps to maintain, Facebook Ireland is nothing without the people who populate it, and the Dublin office has churned out its fair share of movers and shakers.

Colm Long joined in 2009 to establish the EMEA HQ in Dublin, before moving on to run the company's global user operations in Menlo Park. Like Facebook's chief operations officer,

Sheryl Sandberg, Long joined the company from Google, where he held several senior roles involving online sales and operations for emerging EMEA markets.

A Derry native, Long served as director of online operations for the Facebook EMEA HQ and 2009 was meant to be the year this office brought forty jobs to Ireland, but growth of both users and advertising justified extending the investment to add at least seventy jobs and expand key functions such as advertising, finance, legal, product marketing and engineering. Long saw this as a sign of a job well done and recognition of the Dublin office's potential.

Also heading the Facebook fray in 2009 was University of Limerick graduate Shane Crehan, a former financial controller at Dexia and Paddy Power. Crehan was appointed head of international finance at Facebook and, to date, is one of only two listed directors of Facebook Ireland Limited. The second is Sonia Flynn, who was appointed to the board in July 2012. Flynn had been part of the team that established Google's European headquarters in Dublin and came with international-management experience in other multinational technology companies. With this on her résumé, she was the ideal candidate to become managing director of Facebook Ireland.

There were others who worked their way up from the bottom, too. Engineer Colm Doyle joined Facebook in April 2010 from McConnells Advertising, where he had tinkered with Facebook apps. When he attended a Facebook Developer Garage event in Dublin, he impressed the local staff and was offered work at the social network. This was followed in 2012 with an offer to work in Silicon Valley.

Doyle was one of the first members of Facebook's developer-support team, which was responsible for providing guidance to the global community of platform developers, and moved to Menlo Park in 2012 to establish the team's North American presence. Just weeks after his arrival in California, he helped cobble together a last-minute project which leveraged the social network at a key moment in its history. As part of a small team of engineers, Doyle helped rig the NASDAQ bell so that Zuckerberg's Timeline would automatically update the second he rang it, signalling the company's stock market flotation.

Another Irish 'Facebooker' who did well out of the company's IPO is David Spillane – albeit, in a very different way. A former Ernst & Young employee, Spillane was the most senior Irish executive in Facebook's worldwide operation, having joined as chief accounting officer in January 2009. He made headlines following the IPO when he sold more than half of his shares – 256,000 of them – for about €4.25 million.[57] Spillane later left Facebook, tendering his resignation in April 2013. A year later, he became chief accountant at Irish tech success story, Stripe.

Facebook and the Data-Protection Commissioner

One Irish institution hugely affected by Facebook's move to Ireland was the data-protection commissioner (DPC). With a number of online services with worldwide customers establishing European headquarters in his jurisdiction, Billy Hawkes and his team had to police the data protection of citizens around the world, not just in Ireland. By the time Facebook's Dublin move was rumoured in 2008, questions were already being raised about whether the company was compliant with European data-protection legislation – and how the DPC would deal with it if it wasn't.

In 2011, as Facebook reached 800 million users, a DPC audit

was arranged. Facebook had twenty-two complaints lodged against it by lobby group Europe-v-Facebook, and the DPC had to step in and examine all of its activities outside the US and Canada, for which Facebook's EMEA HQ was responsible. Ahead of the audit, Facebook refuted claims from Europe-v-Facebook that its 'Like' button was being used to track people as they browsed online. While the site recorded user information for a period of time, it claimed this data would either be deleted or anonymised after a period of ninety days, and never sold to advertisers.[58]

The DPC investigation began in earnest in October 2011. It was just one of twenty-eight audits undertaken by the DPC that year, but it was the one that garnered by far the most media attention. It was also the DPC's most extensive and resource-intensive review to date, involving up to 25 percent of its staff for its three-month duration and requiring external technical assistance from University College Dublin.

In December, the DPC and Facebook agreed that some best-practice improvements would be implemented over the coming six months, followed by a formal review in the summer. These recommendations included simplified language, an enhanced ability for users to control their own data and greater transparency.

The cultural landscape of the time was one where 84.2 percent of Irish Internet users were using social networks, and 77 percent were sharing personal information, interests, photos and videos on Facebook. Predictably, the realisation that private data could be used by businesses or governments made many of these users uneasy.

Meanwhile, tough new rules were proposed by European Commission vice president Viviane Reding in January 2012, which would penalise companies up to €1 million or 2 percent of annual turnover in the event of a violation of EU data-protection rules.

Facebook missed the end-of-March deadline to get many of its affairs in order for the DPC in 2012, but the commissioner's office was satisfied that progress was steadily being made. Europe-v-Facebook wasn't so forgiving, though, and complained to the European Commission that the Irish DPC was neither sufficiently enforcing the law nor imposing sanctions on Facebook.[59]

The DPC's relationship with Facebook continued to evolve as new products were developed to ensure constant compliance with European laws. Facebook had been hosting enough 'citizens' to make it the third-largest country in the world since 2010 and, in 2012, the social network was hurtling towards a milestone 1 billion active users, who were uploading 300 million photos per day. However, a feature that used facial recognition to scan these images for the faces of connected friends and suggest tags was not received well by those who didn't wish to be identified in this way.

As the DPC began following up on progress made since the audit, it discovered that Facebook had gone beyond the call of duty by switching off its facial-recognition technology for European users. The technology identified people in users' photos and suggested name tags. This had not been requested in the audit's recommendations, but the company decided to put this feature on the back-burner until it found a better way to approach it and educate users on how it worked. Tag suggestions remain disabled in Europe.

As far as the DPC's recommendations were concerned, the follow-up review found that the great majority had been acted upon, and that Facebook had also implemented clear retention periods for the deletion of personal data and enhanced the user's right to have ready access to their personal data. For those recommendations still hanging in the balance, a four-week deadline was set. While this mid-2012 re-audit gave Facebook the all-clear in terms of progress towards data-protection compliance, the social network was ever-changing, and a revelation was coming that would alter forever how users viewed their online data.

In the summer of 2013, IT-professional-turned-whistleblower Edward Snowden leaked classified documents from the US National Security Agency (NSA) uncovering a global surveillance programme allegedly involving a number of government agencies around the world, which was harvesting data from telecommunications and online services.

The NSA revelations sparked privacy concerns regarding data from social networks, and Ireland's DPC was caught in the crossfire. Hawkes determined that companies such as Facebook were permitted to transfer data back to the US-based parent company in compliance with EU law, as long as certain criteria were met under a safe-harbour agreement. Companies that have signed up to these principles agree to ensure EU standards of data protection will continue to be applied even if data is transferred outside its remit.[60]

Evidence on Hawkes's desk suggested that Irish people were marginally affected by the NSA's snooping. Global companies such as Facebook and Google only granted access to user data in response to specific requests for basic subscriber information. Facebook received up to 12,000 of these requests from US authorities in a six-month period in 2013, from the NSA down to local sheriffs. In Ireland, however, data requests from Irish authorities in the same period totalled thirty-four, and not all of these requests were granted.[61]

In light of this, the DPC took the decision not to investigate companies such as Facebook on the transfer of personal data to the NSA. However, the Snowden scandal had highlighted a flaw in the safe-harbour agreement, prompting a European Commission review and high-level talks between the US and EU on the alleged surveillance programme.

Overall, Hawkes's impression of Facebook was one of a company that was keen to comply and to demonstrate that it had done so, but, once again, Europe-v-Facebook was dissatisfied and claimed the DPC had failed to investigate the social network thoroughly in light of Snowden's claims. The group's complaint reached the High Court in Ireland, where it was referred to the European Court of Justice (ECJ). This development was welcomed by Hawkes, who deemed it appropriate for the ECJ to consider this critical issue and re-evaluate the EU law protecting it.

A Cosmopolitan Cluster Forms

In terms of social media in general, there are many big players in the game. But, when you think of professional social networking, there's only one compound word that springs to mind: LinkedIn.

Young entrepreneur-to-be Mark Zuckerberg was still a freshman at Harvard when LinkedIn, a social network for professionals, was founded by a team led by PayPal executive vice president Reid Hoffman. The site launched in May 2003 and became a profitable venture three years later.

LinkedIn has long been claiming the title of the world's largest professional networking site and today supports 300 million members from over 200 countries. As users flocked to the site that would help them to build a professional profile online, LinkedIn expanded with offices in London and Amsterdam – but it was in Dublin that the company decided to locate its international headquarters.

In March 2010, LinkedIn announced its decision to put its international headquarters in Ireland's capital, to support its expansion in Europe and further afield. At the time, the site counted 60 million members, 14 million of whom were in Europe. Dublin would serve as the centre of the Californian company's international growth and recruitment began immediately for staff in marketing, sales, finance and customer service.

By the summer, finance roles became a key focus: thirty-five positions for finance professionals were announced in June. Sharon McCooey, a graduate of University College Dublin and Dublin Business School, and the freshly appointed international finance director, was tasked with heading up the operation.

Financially, the company was doing well. It had reached the point where it was making acquisitions, starting with Mspoke in 2010. By the close of the year, LinkedIn was valued at over $1.5 billion.[62] Following this valuation, LinkedIn filed for an initial public offering in January 2011. Before the NYSE ticker tape read LNKD, the company celebrated its first year in Dublin with the announcement of a hundred new jobs.

Roles in HR, business development and operations were added to the mix, with recruitment set to take place over the coming year. At the time, international HR director Connie Gibney cited talent, international language skills and the workforce's experience in fast-growing Internet companies as key to LinkedIn's decision to expand in Dublin. In its first year, the Dublin office had grown to a team of seventy, and the network was seeing new members sign up at a rate faster than one per second.

That summer, LinkedIn traded its first shares on the New York Stock Exchange at $45 each. At their peak, shares rose 172 percent in this first day of trading, closing 109 percent above the IPO price at $94.25. Later that year, Gibney was happy with how recruitment was going at the Dublin HQ. For a site supporting nine languages, hiring multilingual staff was key and Dublin's diverse community was delivering. Following years of foreign direct investment, the Irish capital had attracted staff from all corners of the globe, and many students came to Ireland to study English and remained for work. Hiring staff with the required language skills – whether Russian or Turkish – proved relatively easy, with only Dutch-speaking staff evading LinkedIn's recruiters.[63]

Being an English-speaking nation housing a multicultural workforce continued to be a winning formula for Ireland. International headquarters need international staff, after all. This is just what was achieved in the establishment in Dublin of Twitter's EMEA HQ – the largest Twitter office outside the US – where about half the staff are Irish and the other half are international.

Ireland Is #trending

Twitter's growth was, in a word, rapid. On 21 March 2006, New York University undergraduate student Jack Dorsey sent the first ever tweet: 'just setting up my twttr'. Since then, the 140-characters-or-less micro-blogging service has added some vowels to its name, built a revenue model around advertising and signed up more than 250 million users worldwide.

Twitter's popularity spurt was braced by a successful promotional play at South by Southwest Interactive (SXSWi) in Austin, Texas during the spring of 2007. The tech conference contingent of the multimedia festival marked a tipping point for the burgeoning social network as usage increased from 20,000 tweets per day to 60,000 thanks to two cleverly placed sixty-inch plasma screens in the conference hallways dedicated to streaming visitors' tweets.

Fast-forward to June 2010, and the site was supporting 65 million tweets per day, which equates to about 750 tweets per second. Tweeting records were broken by major sporting tournaments, from the FIFA World Cup in South Africa to the NBA Finals in Los Angeles.

By March 2011, 140 million tweets were being sent daily and, in September of the same year, Twitter opened its European headquarters in Dublin. The announcement was made by IDA Ireland, naturally, in a tweet. 'Ireland is trending,' the @IDAIRELAND Twitter account proclaimed. 'Twitter to establish international office in Dublin.' Topped off with an #idairl hashtag, Twitter had arrived in Dublin amid a fanfare it had itself established. At this point, the network was supporting over 100 million active users generating over 230 million tweets per day.

In the summer of 2012, at more than six years old, the company shook off its downy fledgling logo and emerged as a new-look Twitter bird that was sleek, confident and pushing forward. The 200-million-monthly-active-user milestone was passed by the close of the year after Twitter announced its acquisition of Vine, a video-sharing platform that, like its new parent company, encouraged creativity through limitations. The six-second-video publisher was launched as a standalone app in January 2013.

Twitter preceded its preliminary S-1 filing with the US Securities and Exchange Commission in October 2013 – the first step on its road to an IPO – with a doubling of its team in Dublin. Staff numbers had surpassed a hundred and plans were revealed to have a hundred more by the end of 2014. To accommodate this growth, Twitter Dublin flew the nest it shared at Fitzwilliam Hall and set up shop in The Academy on Pearse Street.

After arriving for his first day in the new office, Twitter Ireland's managing director, Stephen McIntyre, tweeted about the company's new Pearse Street home and asked for a local cafe recommendation. Within minutes he received reply from Cup Cafe on Nassau Street, welcoming Twitter Dublin to the neighbourhood with a walking map to their doorstep.

As 2014 dawned, the Dublin office had its own 140 characters in situ and was servicing more than fifteen business functions, including engineering at its Command Centre, where the integrity of the real-time platform is maintained on a large scale. But despite all the hard work, fun was being had at Twitter Dublin too, with senior staff conducting all business meetings in costume one Halloween.

Ireland Gains New Friends

Social media giants such as Facebook, Twitter and LinkedIn have reshaped the way we see the world and communicate with the people around us. But they have also helped reshape Dublin's docklands.

Facebook updated its status to hundreds of jobs in Dublin, LinkedIn became a very important connection and @Twitter got #bigger. Between them, the three firms now employ more than a thousand people in the Silicon Docks area.

With the capital a home away from home for Facebook, Twitter and LinkedIn, it would only be a matter of time before their Silicon Valley friends, followers and connections came knocking.

Grand Canal Docks and Sir John Rogerson's Quay, 1926
Courtesy of Dublin City Public Libraries and Archive

The gasometer on Sir John Rogerson's Quay (right) in the early 1970s
Courtesy of Dublin City Public Libraries and Archive

The Samuel Beckett Bridge and Sir John Rogerson's Quay, with the Poolbeg chimneys in the background. Photo: Kristina Petersone, 2014

Grand Canal Basin with Accenture's offices, the Bord Gáis Energy Theatre and the Marker Hotel. Photo: Kristina Petersone, 2014

Grand Canal Square, designed by American landscape architect Martha Schwartz, opened in June 2007. Photo: Kristina Petersone, 2014

Boland's Mill on Grand Canal Street. The building was occupied by Eamon de Valera during the 1916 rising. Photo: Kristina Petersone, 2014

Google's first Irish employees pictured with the search engine's founders Larry Page and Sergey Brin in 2003, in the company's then office in Seagrave House on Earlsfort Terrace. Photo courtesy of Oisín Ó Mír

Google's EMEA headquarters on Barrow Street. Photo: Paulo Cabbarao, 2014

Facebook's home at Grand Canal Square. Photo: Kristina Petersone, 2014

Facebook chief operating officer Sheryl Sandberg at the social network's Dublin office in April 2014. Photo courtesy of Facebook

Inside the offices of Airbnb in Dublin. Photo: Ed Reeve

Whiskey bottles and a keg adorn a themed, traditional Irish bar inside the Dublin offices of Airbnb. Photo: Ed Reeve

Web Summit founder Paddy Cosgrave, Taoseach Enda Kenny and NASDAQ executive vice president Bruce Aust join others on stage for the ringing of the NASDAQ opening bell at the 2013 Web Summit. Photo courtesy of Web Summit

Paypal and Tesla co-founder Elon Musk onstage with Taoiseach Enda Kenny at the 2013 Web Summit. Photo courtesy of Web Summit

Squarespace chief operating officer Jesse Hertzberg with Taoiseach Enda Kenny. Photo: Jason Clark

Twitter chief executive officer Dick Costolo pictured with staff at the company's Dublin office. Photo courtesy of Twitter Dublin

6. Internet Influx: The Rest Start Arriving

Elaine Burke

Following in Facebook's footsteps, high-growth tech start-ups began flocking to Dublin to create European hubs, support teams and development centres. Often announcing their arrivals with triple-figure jobs announcements, these companies were welcomed with open arms by Ireland's ailing economy. Before long, ghostly vacant properties along Dublin's waterways were teeming with life and recovery was in sight.

IDA Inspires Zynga

In March 2010, while the IDA was welcoming LinkedIn to the south docklands, Zynga Game Ireland Limited was established with a registered address at the office of its legal advisors A&L Goodbody on North Wall Quay, across the river. Word leaked out, and the social-gaming developer's plans to build a base in Dublin were soon confirmed by spokeswoman Shernaz Daver.[64]

In the year that followed, San Francisco-headquartered Zynga made a number of acquisitions, establishing new bases in Boston, New York, Austin, Dallas, McKinney (Texas) and Frankfurt, Germany. It wasn't until summer 2011 that the official opening of Zynga's biggest European office finally came, eschewing the growing trend of setting up by Grand Canal Dock with an office in The Oval building in Ballsbridge, a little bit further south-east.

Zynga had an early start in registering its Irish operation, no doubt because the IDA had been early to make contact, in mid-2009, before the fast-growing games developer was on any other agency's roadmap. Chief operations officer Marcus Segal said the IDA partnered with Zynga in the early days and did a great job of showing them the way – the way to Dublin, that is.[65]

By Segal's recollection, the IDA proved invaluable in Zynga's plans to locate its European headquarters in Ireland. The Irish agency visited the US firm early and often and, by the time others came knocking, Zynga was already bosom buddies with those who had been there from the beginning.

Speaking with Web Summit founder Paddy Cosgrave at a US Embassy conference in Dublin in April 2014, Segal – who at this point had departed Zynga to become the Web Summit's first entrepreneur-in-residence – said his experience with the US company's Irish operation was such that he didn't think it possible to be a major global player without one. As well as being first on the suitors list, IDA Ireland made it easy for Zynga to get started in Dublin. It took about six months to set up the Irish operation – a timeframe Segal didn't believe possible even in another US state.[66]

In July 2011, following the celebration of Zynga's official opening in Dublin, the company filed for what was expected to be a $1-billion IPO. Its trajectory since hasn't been that of Facebook, LinkedIn or Twitter. In fact, Zynga ended 2012 on a list of CNN's top ten tech fails of the year, and staff cuts and stringent times were to follow.

However, Zynga's move to Ireland was still a win for the Irish government in 2011, and newly appointed Minister for Jobs, Enterprise and Innovation Richard Bruton, was wholly committed to aggressively targeting expansion in innovative high-tech sectors to achieve growth in the economy and get the country back to work. He was intent on making more announcements of the same ilk in the months and years to follow and, indeed, he soon began to repeat himself with jobs announcement after jobs announcement powered by foreign direct investment.

Where It All Began

Fitzwilliam Hall, just shy of being on the bank of Grand Canal in Dublin 2, was where Facebook first began to germinate in its early Dublin days. It was in this serviced office suite that the team grew from two to fifty before it had to be replanted to a more spacious docklands location. Twitter's international headquarters also started off in this building in 2011, and so it was a natural fit for Dropbox, when that company was making its first move outside of the US in December 2012.

The cloud-storage provider had already reached over 100 million users in 200 countries – including ten users in Antarctica – and it was serving all of them from its sole office, in San Francisco. More than one-third of its users were based in Europe and CEO and co-founder Drew Houston was keen to be closer to these customers through an international hub.

U2 rockers Bono and The Edge are among Dropbox's investors and the famous Irish frontman was confident the company would find the smart and innovative workforce it needed in Ireland. In his opinion, the IDA played a blinder in securing Dropbox's commitment to Dublin.[67]

The support created in Dublin enabled Dropbox to have sales teams available at nearly all hours of the day as well as customer service in added languages. To emphasise the development of its multilingual support, the announcement appeared on the Dropbox blog in languages from Spanish and German to Japanese and Korean – and even *as Gaeilge.*

When Dropbox eventually outgrew its digs in Fitzwilliam Hall, instead of moving closer to Grand Canal Dock, the company edged further away – to Park Place on Hatch Street, in the heart of the south city centre. Development was also spreading across the Liffey, with HubSpot setting up on North Wall Quay and TripAdvisor following with an engineering hub in the Liffey Trust Centre on Sheriff Street.

HubSpot first announced its intentions to come to Dublin in September 2012 along with its mission to bring inbound marketing to Europe. Following phenomenal 200-percent annual growth in its international business segment, the software-as-a-service (SaaS) firm planned to hire 150 people at its first branch office. The timing coincided with the launch of HubSpot 3, which came with the ability to translate any landing page into any language and to localise SEO tools with data and keyword suggestions from the native country. HubSpot was ready to conquer an international market and intent on becoming a major employer in Dublin, growing to 250 people within three to four years and from 300 customers in Europe to several thousand over the same period.

Chief operations officer J. D. Sherman felt a Silicon Valley vibe about Dublin, which he comfortably referred to as the digital hub of Europe. CEO Brian Halligan admitted that tax incentives were a strong draw (the marketing mogul was honest enough not to add some spin to that), but he asserted that the talent available was also enticing.[68]

In March 2013, Halligan was keen to express that the Dublin office was much more than just a call centre offering customer support. According to him, this office was envisaged as a mirror to HubSpot's US headquarters in Cambridge, Massachusetts – a DubSpot, if you will – with core engineering and R&D happening alongside sales and business operations.[69]

TripAdvisor was similarly encouraged to leverage Dublin as more than just a customer-support centre. In September 2013, the announcement came that the thirteen-year-old tech firm would establish its largest engineering hub outside of the US in the Dublin docklands area. Recruitment began immediately for at least fifty core engineering roles, with site manager Lars Holzman eagerly putting himself out there at events such as Career Zoo in the Convention Centre Dublin to meet potential candidates face-to-face and encourage them to join the team.

Holzman later became director of engineering at TripAdvisor Ireland and his team grew faster than expected, with ten members of staff already recruited after its first quarter in Dublin. Andy Gelfond, senior vice president of engineering and operations, said on a visit to Dublin that staff in Dublin were already contributing to major project initiatives, including playing a significant role in a vital change for the service: the ability to make bookings from the mobile site. One recruit in Dublin was also contributing to analytics and personalisation, another high priority for development on TripAdvisor's 2014 roadmap.

The Digital Sprawl Continues

As Dublin's digital cluster began to sprawl outwards, the heart of the docklands region continued to attract its fair share of tech talent. The Bloodstone Building long stood as an anachronism on the River Liffey's banks during the recession; a landmark high-rise office space built to cater for a groundswell of FDI just a few years ahead of its time. The property eventually fell to the National Asset Management Agency (Nama), but things began looking up in 2012 when LogMeIn joined SEB Life International, a subsidiary of Nordic financial services group SEB, in the Seán Dunne-developed property, officially announcing its arrival in November of that year.

LogMeIn, headquartered in Boston, is a cloud-based software-as-a-service company providing remote access products to clients. A month earlier, another cloud-based player, Zendesk, had announced plans to establish a Dublin centre.

Starting with just two engineers in its initial base in the Fitzwilliam Business Centre on Sir John Rogerson's Quay, Zendesk planned to grow its Dublin office to twenty-five by the end of 2013. This number was reached a month ahead of time and the company's headcount more than doubled to fifty-two over the first six months of 2014. It was then that the firm's new office overlooking the Grand Canal was officially opened.

One Grand Parade, almost 3,000 square metres across six storeys, had the unfortunate luck of being completed at the height of the recession in 2009 and, naturally, saw difficulties in finding residents. However, by spring 2014, five floors were occupied in the energy-efficient building – three of them by Zendesk. Dublin-based vice president of product engineering Colum Twomey said the new office had capacity for 150 people, allowing lots of room to grow for a company providing customer service in over 140 countries and forty languages.

The official opening of Zendesk's Dublin development centre at One Grand Parade was an encouraging sign. IDA client companies were moving in and filling the empty spaces that had previously, in their ghostly vacant states, served as haunting reminders of the country's foolhardy ride atop a Celtic Tiger economy before being unceremoniously dismounted.

In 2013, Bloodstone developer Seán Dunne was filing for bankruptcy in the US after leaving Dublin and the property crisis behind, but times were changing in the Irish capital. Yet another EMEA HQ arrived in Dublin's docklands, initially at Dogpatch Labs on Barrow Street and then at the Fitzwilliam Business Centre.

Dublin was selected as Squarespace's first location outside of New York City in order to tap into its friendly, young and talented workforce, according to founder Anthony Casalena. Casalena and the Squarespace team had first been contacted by the IDA almost twelve months previously, at their headquarters in New York, and, in the time that followed, they had received assistance in plans to open an Irish office.

Another SaaS company, Squarespace provides individuals and businesses with the tools to easily build and maintain websites and blogs. One-fifth of the company's business was located outside the US and customer service needed to be built up to ensure 24/7 global support – a key selling point of the Squarespace brand. The Irish workforce were seen by Squarespace as 'customer service people': likeable and friendly.[70] It thus made perfect sense for the company to locate in Dublin and tap into that talent.

Service with a Smile

By now, Dublin's reputation as a customer-service centre of excellence could have been written in stone, thanks to the number of established and emerging tech companies running global support out of the Irish capital. To attribute this solely to like attracting like would be to ignore the inexplicable magnetism of that charming collective known the world over: the Irish people.

As the country's tourism organisations will gladly tell you, Ireland has a glowing reputation as a friendly nation. Indeed, the

National Tourism Development Authority takes its name from the Irish word for 'welcome': Fáilte Ireland. Naturally, the land of a hundred thousand welcomes would be the obvious choice for Airbnb's Hospitality Innovation Lab.

In 2007, Brian Chesky and Joe Gebbia were struggling to pay rent on their San Francisco apartment but, with a design conference coming to town and increased demand for short-term accommodation, they had an idea. The roommates turned their living room into a B&B, offering up to three guests an airbed to sleep on and a homemade breakfast in the morning.

The ingenious idea didn't stop there. In spring 2008, Chesky and Gebbia enlisted the help of software engineer Nathan Blecharczyk to build Airbedandbreakfast.com, which launched in August of that year. What later became known simply as Airbnb continued operating out of that same three-bedroom San Francisco apartment even as more staff were added. It got to the point where Chesky sacrificed his bedroom for the cause and decided to live through Airbnb accommodation.

In November 2010, the company's rapid growth attracted $7.2 million in Series A Funding from Greylock Partners, Sequoia Capital and others. A milestone was reached in February 2011, when the site celebrated 1 million bookings. Airbnb was quickly becoming a buzzword on the tech scene.

Through events such as SXSWi, where gatherings of the tech-savvy need to bed down for the night, it was growing its community. By July 2011, it was named in the same breath as Dropbox and Quora as the next-generation of multi-billion-dollar start-ups by the *New York Times* and, pretty soon, it became common for new concepts to be described as 'the Airbnb of . . .' whatever industry they hoped to disrupt.[71]

As 2012 began, 5 million nights had been booked through the service, which doubled to 10 million nights by June of that year.

The company also expanded internationally throughout the year with offices in Hamburg, Paris, London, Milan, Barcelona, Copenhagen, Moscow, São Paulo, Sydney, Singapore and Delhi.

It was in July 2013 that the start-up first began making moves towards establishing home in Dublin. Speculation around an Irish office was growing. The first concrete evidence of Airbnb's decision to locate in Dublin came not with a noisy announcement, but with four positions quietly advertised on the company's website. A core team was established by August and began operating out of an office on Lansdowne Road. September came with the official announcement of the opening of its European hub in Dublin, and the promise of turning this location into a Hospitality Innovation Lab.

Ireland, the land of *céad míle fáilte*, has hospitality in its DNA. This reputation attracts something that businesses such as Airbnb need: international talent. Dublin is home to a multicultural population and Chesky saw it as an emerging technology epicentre in Europe, as well as one of its most international cities, with a wide range of languages represented. Airbnb was preparing to tap into what Twitter, LinkedIn and others had already discovered.

From its foundation in 2008, Airbnb took about four years to reach 4 million guests served. Less than a year later, in October 2013, it had served 9 million. It seems early predictions that this start-up would be a unicorn – a billion-dollar company – fell far short of the mark. In April 2014, the company was valued at $10 billion, following $475 million in Series D investment led by private-equity firm TPG Capital.

To put this in perspective, at that time, fast-growing social network Pinterest was valued at $3.8 billion, music-streaming service Spotify at $4 billion and rocket-makers SpaceX at $4.8 billion. The young disruptor borne of a San Francisco apartment was leaving these breakthrough start-ups in the dust.

As plans for Airbnb's Dublin innovation hub began to see

fruition, a hundred new jobs were created in order to double the size of the team. These roles focused on customer service and safety, while long-term plans included pan-European roles in HR, finance, technology and legal. Indeed, Airbnb's founders envisaged that all of the company's functions could one day operate out of its Dublin base.

EMEA customer experience head Aisling Hassell leads the 100-strong team at the Watermarque Building in Ringsend, a uniquely designed office space reflecting locations listed on the site. Hassell has declared a three-fold purpose for the Dublin office. As well as serving as a Hospitality Innovation Lab, she sees it as the centre of Airbnb's customer service operations, and the point of contact between San Francisco and its other international operations.

While Airbnb's founders have not yet earmarked Dublin as a development hub for its technology, it is certainly a key cog in the company's customer-service machine, which is a major part of its business. Being the world's fastest-growing accommodation network without owning a single property can only happen with a strong, stable, safe and well-supported community of guests and hosts. In Dublin, it is the job of hospitality innovator Clément Marcelet to ensure these hosts are the best that they can be by running meet-ups, webinars and training sessions and hosting other educational resources online.

Engine Yard Expands into Ireland

Searching for the seemingly magnetic force attracting tech-based FDI to Dublin's docklands, we can see purpose-built-yet-unoccupied buildings being sold for a song through Nama, or tax incentives and silver-tongued IDA agents buttering up Silicon Valley CEOs. But we can also see the people: the Irish and the adopted Irish.

On the first day of the Web Summit in late October 2011, Engine Yard, the platform of choice for building cloud-based applications, officially announced that it would establish its EMEA headquarters in Dublin. Though an IDA client company, Engine Yard wasn't necessarily drawn to Ireland solely for tax benefits or the opportunity to be Google's new neighbour on Barrow Street.

A software developer since the 1990s, Eamon Leonard was approaching his thirtieth year when, in 2007, he decided to quit an unfulfilling job to go freelance. A year later, he was introduced to David Coallier through a mutual friend on Twitter. Coallier, a Canadian, had romantic notions about living in Ireland, inspired by his Irish grandparents. The two met for a pint when Coallier was in Ireland to speak at a conference and, halfway through their conversation, Leonard asked Coallier if he wanted to start a company.

Coallier moved to Ireland within months, settling in Cork, and the pair established a consultancy under the name Echolibre. Leonard and Coallier quickly realised that they only wanted to work with start-ups and, over the next two-and-a-half years, they helped thirty of them from Ireland and abroad.

The duo had learned a lot but, at this point, they were itching to make products of their own. They teamed up with Iceland native Helgi Thorbjoernsson in March 2009 and, in conjunction with coder Iarfhlaith Kelly, developed a URL shortener – Short.ie – which was popular among Irish Twitter users. However, when Bitly raised millions of dollars in funding, the Ireland-based team gave up in the face of powerful competition from the US.

The Echolibre team were well aware of San Francisco's Engine Yard and its platform, which helped minimise the tedium of repetitive tasks required in development using the Ruby programming language. They also realised that a similar service based on PHP would plug a gap in this market and, in early 2011, Orchestra was born. Negotiations with venture capitalists began almost immediately but, by the end of April, a developer from Engine Yard had reached out, eager to partner up.

Orchestra had developed a platform for deploying, scaling and managing PHP applications – something Engine Yard needed in its arsenal. In addition, the US company wanted to expand into Europe and saw Dublin as the perfect base. Meet-ups with Orchestra followed, including a visit from CEO John Dillon, and, in August 2011 – one week after Leonard's honeymoon – the deal was signed.

Engine Yard subsumed Orchestra and announced the establishment of its EMEA headquarters in Dublin in October 2011. In joining forces with Leonard's small team, the platform-as-a-service strengthened its technological and geographical coverage. Leonard became vice president of engineering and put the Dublin office to good use as a spot for software gatherings, which attracted experts from around Ireland and Europe. Engine Yard supplied the space (and beer and pizza) required, and Leonard brought the community.

To date, the Dublin PubStandards events run by Leonard and supported by Engine Yard are the largest monthly gatherings of developers, designers and start-ups in Ireland, attracting up to 160 techie people to the Bull & Castle bar by Christ Church Cathedral. Leonard is also an angel investor with Irish start-ups Intercom, Trustev, Rentview, PicStash and Brewbot in his portfolio and, in the spare time that he somehow manages to find, he provides advice to other start-ups hoping to reach similar levels of success.

Dublin's Got Talent

Throughout this chapter, the IDA's tenacity in chasing down high-growth tech start-ups from an early stage has been well documented and its successes are told by the numbers. In each case, proclamations that the talent and multiculturalism offered by Dublin edged out the myriad other cities undoubtedly pitching for investment are certainly plausible, yet these words are like a thin veil disguising the deals taking place on countless IDA business trips and the economic incentives promised over each handshake.

In the case of Engine Yard, we see a different approach: a US company coming to Ireland explicitly to connect with technology that was created there. The Engine Yard win made a champion of entrepreneur Eamon Leonard, who epitomises the very reputation the IDA so fervently advertises to potential investors. In his various guises as visionary entrepreneur, community builder, education advocate and investor, Leonard's attitude perfectly encapsulates Dublin's attraction beyond tax breaks and government promises

Therein we have the nugget of undeniable truth in repetitive statements that talent is key to Dublin's huge FDI success, and nugget by nugget forms a cluster deserving of a title – Silicon Docks – implicitly comparing it to one of the world's greatest tech hubs.

7.
A Tech Ecosystem Develops

Philip Connolly

Luca Boschin and Alessandro Prest had heard about Dublin so, in 2014, they came to see the place for themselves. The Italian duo had originally set up in Switzerland. When it came to moving on, Dublin beckoned as the best destination to get their tech firm LogoGrab off the ground.

Only a few days off a plane in June 2014, Boschin and Prest are well into a crash course of one of the quirks of the city's budding start-up scene – Silicon Valley may have its cafes and restaurants, but Dublin has its pubs. In this case, the pub is The Stag's Head, located off Dame Street in Dublin's city centre. It is hosting Techpreneurs, one of the city's original tech meet-ups.

The event is running a little late and duo are wondering what the evening will be all about. There will be speeches, they are reliably informed but, for now, they can just relax and have a couple of drinks with their new Dublin comrades. And they need not worry about money –Bank of Ireland is picking up the tab. It too has heard about Dublin's start-up set and wants to get in on the action.

Tonight, at the invite-only event, the crowd is a mix of heavy hitters and young hopefuls. Barely in the room a few minutes, Boschin and Prest are already handing out business cards and pitching LogoGrab, which professes to be planning to kill the QR code with its ability to scan any logo from a smartphone and link it to a webpage.

The duo have already caught the attention of Gary Leyden, one of the most recognisable faces among Irish start-ups. Leyden runs Launchpad, an accelerator programme that is responsible for a huge number of Ireland-based start-ups that are traversing the globe in search of sales and investments.

They have also been quick off the mark in terms of setting up an office. Rather than looking to get in on Dublin's booming office market, the duo have taken a more innovative approach. Just across from Silicon Docks at the IFSC, in the part of town where billion-dollar hedge funds trump tech giants, Boschin and Prest have rented a duplex apartment.

The bottom floor will be for sleeping, the top converted into an office for work. Now, in one of the many apartments built during Ireland's feverish property boom that was designed to appease lavish Celtic Tiger cubs, it is the tech elite who reign. The duo have inherited some of the best views of the city, which can be viewed from their very own terrace pool.

They have already raised close to $1 million, some of which came from Enterprise Ireland, and now plan to get going on sales. Aside from the intimacy of Dublin, the fact that half of their potential customers are now on their doorstep was another big selling point. For it isn't just Irish start-ups that are looking to thrive in Dublin's tech world, Europe has taken notice too.

The talk of the evening, however, will give the two men a window into some of the more-prescient debates that are taking place among Ireland's tech elite. While the title of the evening is 'Start-up Scene versus Software Industry', the talk of the evening has really surrounded the question of where exactly Ireland's start-up scene is going.

It is still young, but it is no longer a new phenomenon. However, it still has not seen a real hit. There have been million-euro funding rounds, and more than a few exits (in which founders have left their companies, either by selling them or through IPOs), but no Irish start-up has really hit the big time.

The question of where to go next splits the audience. For some, Dublin needs to figure out what its *raison d'être* is, what exactly it offers beyond low taxes, talent and questionable weather. For others, a massive exit and the validation that comes with that is just around the corner. It is a debate that has become a refrain among Irish founders.

The Rise of Accelerators and Incubators

Beyond simply being able to have a pint with a potential mentor or investor, never before have there been so many supports for budding entrepreneurs, from accelerator programmes to incubators and venture capital.

While estimates differ, the common consensus is that around 2009 something changed. Against the backdrop of impending market turmoil, as Ireland lurched from one financial meltdown to the next, somewhere in Dublin a spark ignited a new wave of Irish businesses.

It was also in 2009, that the National Digital Research Centre (NDRC) established its Launchpad accelerator programme, and it is here that many of Ireland's brightest start-ups came into being. While the concept of start-up incubators and accelerators dates back to the 1950s, when US companies gave entrepreneurs with little more than an idea some office space and mentorship in exchange for equity or a fee, a new model emerged in 2005.

That year, American seed accelerator Y Combinator rewrote the manual on how to develop start-ups. In exchange for equity, the Y Combinator accelerator gave start-ups money, advice, practical training and office space as well as introductions to potential investors, partners and clients. Two billion-dollar companies later – Dropbox and Airbnb – and few are arguing with the model.

Before Launchpad, little in the way of support of that type existed in Ireland. The NDRC, which was primarily focused on commercialising research, looked internationally at what was working. Across the Atlantic, it found the answer with Y Combinator. Launchpad took that model and, in the spirit of any good start-up, has evolved over the past five years into very much its own setup.

For Gary Leyden, director of the NDRC's Launchpad programme, the idea is to see if what a company is building will work and, more importantly, if anyone is willing to pay for it. If not, then the company can use market research to 'pivot' – to rejig the offering or target a completely different market – or they can scrap the idea and go back to the drawing board. He says one of the most important impacts that the Launchpad programme has had over the past five years is to change the culture around setting up a firm, and the type of people that it produces.

The programme promotes a type of thinking that is as valuable to existing companies as the start-up ecosystem. For someone to try a start-up and then go to work for an existing company brings a completely different mindset, Leyden says. The multinational tech giants helped to create the start-up ecosystem and now that ecosystem is starting to produce their ideal employees – the system, it seems, is starting to perpetuate itself.

In 2014, the NDRC was ranked in the top 2.5 percent of incubators worldwide, and proof of the Launchpad programme's success can be seen in the rise of its graduates such as Soundwave and Logentries.[72]

Within a month of its launch in July 2013, Soundwave's music-discovery app had been downloaded more than 200,000 times in

190 countries. It has since racked up more than 1 million downloads, and was even endorsed by Apple co-founder Steve Wozniak, who declared it 'a music product that so perfectly fits my life'. The company has been backed with more than $1.5 million of venture capital from international investors such as Dallas Mavericks owner Mark Cuban and music-industry heavyweight Trevor Bowen, one of U2's business advisors.

Logentries, founded by Trevor Parsons and Viliam Holub in 2010, went through NDRC's Launchpad before moving on to Dogpatch Labs. The company, which provides log management and analytics via a system which Parsons describes as a kind of CCTV for software, has received backing to the tune of $11 million from leading US and Irish investors. It has also built up tens of thousands of users in over a hundred countries.

The NDRC's Launchpad programme has since been joined by accelerator and incubator programmes such as Dogpatch Labs and Wayra in Silicon Docks, and Propeller run by the DCU Ryan Academy across the river at the Liffey Trust Centre.

Dogpatch Labs sits only a couple of doors down from Google's offices on Barrow Street. Set up by US investment firm Polaris Partners as a hothouse for high-growth firms, more than thirty companies have passed through the incubator's doors since it opened in 2011. Among them are Logentries, Boxever and Intercom, which have since raised close to €30 million in venture-capital funding between them.

It was also in Dogpatch that James Whelton spent twelve months incubating CoderDojo. More than 16,000 kids around the world now learn to write code every week as a result of the CoderDojo movement. PizzaBot, a game developed by Cork teenager Harry Moran, who learned how to code at the computer club, knocked Angry Birds and Call of Duty off the top of Apple's app store charts in November 2011.

A New Start-up Hub Emerges

An accelerator programme is all well and good, but what happens once it is finished? That was the position Conker founder Russell Banks and fellow Wayra graduates found themselves in during 2013.

For most start-ups, the key is to stay lean and nimble – long-term office leases and the responsibilities and overheads that come with them are a big turn-off. Rather than going out on their own, the Wayra cohort decided to do something a bit different.

After spending three months together, the companies involved in the programme were not ready to part, and, led by Banks, they banded together and inadvertently set up their very own start-up hub in the Liffey Trust building in the north docklands.

Since then, a number of accelerator graduates have joined them and requests from other start-ups have started flooding in. The rent is cheap, the commitment is short, the space is flexible and the broadband is fast.

The Influence of the Multinationals

As more and more US tech firms have set up in and around Grand Canal Dock, a start-up scene has flourished in the area. In the face of growing youth unemployment and emigration, the influx of Silicon Valley firms injected an energy and innovative edge that gave graduates a different worldview than that of finance- and property-focused Celtic Tiger Ireland.

From the outside, it appears that Dublin is aggressively focused on putting together all the ingredients necessary to create a world-class ecosystem. But so are most other European cities. For investor Barry Maloney, Irish start-ups have a vital trump card – the company they keep in Silicon Docks. Without the shining examples of Google and Facebook, among a plethora of others, Maloney believes that Ireland's start-up scene would barely be on the map.

Maloney is about as close as Europe gets to one of Silicon Valley's infamous money men. His investment firm, Balderton Capital, has backed companies such as LoveFilm, Bebo, Betfair and Wonga. The company has more than $2 billion in funds under management and the start-ups it has backed in its fourteen years of operation are today worth more than $10 billion.

While Maloney may be based out of London, Dublin is home. And looking at what could make Ireland stand out in an increasingly competitive European ecosystem, Maloney can't see much past the impact that firms such as Facebook and Google have had on its nascent start-up scene.

Maloney recalls a meeting he was at in 10 Downing Street at the behest of UK prime minister David Cameron several years ago. The discussion focussed on how London could become a tech hub. He remembers thinking that the city had a long way to go. At that stage, Google, Twitter and Facebook were already Dublin, having followed the original group of firms such as Intel and Apple to Ireland. That is a massive advantage that no one can replicate.

For Maloney, the firms are vital finishing schools for the talent that will carry Ireland's start-up ecosystem on their shoulders. He doesn't believe the ecosystem would ever have started without the multinationals.

The Role of Venture Capital

While the factors that have contributed to a boom in Irish tech may be numerous, it often comes back to cold, hard cash. Here, more than any other area, is where Ireland's start-up scene is a very different place.

Colm Lyon has seen the dramatic change in Ireland's venture

capital environment. In 2000, he set up his online-payments company and went looking for venture investment. It would never arrive. It was not for a lack of trying, but the few investors that inhabited Ireland's early start-up scene wouldn't touch the company that became Realex Payments. Located at the heart of Silicon Docks, it now employs more than 170 people in three countries and has turnover in excess of €18 million a year.

If he were to set up Realex now, Lyon would have venture capital firms throwing money at him. In creating an ecosystem of technology companies, Ireland has attracted some of the world's biggest venture backers to set up operations here, bringing significant capital with them. Blue-chip venture-capital names such as Bessemer, Polaris and Silicon Valley Bank have all bought into Dublin-based firms.

For Will Prendergast, a partner at Ireland's newest indigenous fund, Frontline Ventures, a lot of the change in the ecosystem has come down to how young Irish business people now perceive risk.

While funds did come to Ireland during the dot-com bubble and stayed until the mid-2000s, he recalls the years between 2006 and 2009 as a time when international investors did not show much interest in the Ireland. That has now vastly changed, with the gradual entrance of US funds. Some of that is down to Enterprise Ireland and some is down to the companies that have been created in Dublin. Now that funding ecosystem is starting to settle.

Prendergast says funds often start out as very generalist and risk averse. Over time, they become more specialised and more familiar with their sector and will take on more risk. That is beginning to happen in Ireland with healthcare-only funds, and software-only funds such as Frontline.

Among the companies which Frontline has invested in is CurrencyFair. The peer-to-peer lending platform raised almost €2 million in a funding round led by the VC firm, with a number of other angel investors making up the rest of the investment. A native of Australia, CurrencyFair founder Brett Meyers didn't come to Dublin for the tech scene but for the dockland's other massive international industry – finance. As a former employee of Zurich Bank and JP Morgan, Meyers had plenty of experience in banking and saw a big blind spot.

Like many start-ups, CurrencyFair's genesis came out of frustration. As an expat living in Dublin, Meyers had several bad experiences dealing with international money transfers, racking up hundreds of euro in bank fees. He decided to exchange funds directly with friends whenever he could to avoid the hefty margin – typically 3 percent to 5 percent – built into bank exchange rates and international transfer fees.

CurrencyFair operates on the same principle, but takes away the hassle of having to find a friend. The company has set up a safe and regulated online marketplace, where users exchange currencies with others around the world, anonymously, with the funds being held at all times by CurrencyFair itself, in segregated client accounts, removing risk. Founded by Meyers with Jonathan Potter, David Christian and Sean Barrett, CurrencyFair went live at the start of May 2010 and has now processed more than €600 million worth of transactions.

The Price of Development

The development of Dublin's start-up ecosystem hasn't been free, with significant investment coming directly from state coffers. Over the past decade, Irish VCs have invested hundreds of millions in Irish firms and brought in a similar amount from international VCs through syndication, but Enterprise Ireland also directly spends millions investing in companies through its various funding programmes and venture partnerships.

One of its biggest is the High-Potential Start-up Programme (HPSU), where it puts its money where its mouth is by investing up to about €250,000 in over ninety companies. With many SMEs finding themselves cash-strapped because of a lack of bank lending, a number of lobby groups have called on the government to give them more support. However, the HPSU programme and most of Enterprise Ireland's other funds are not meant to bridge that gap; the qualifying criteria are tough and often require additional venture capital, which lends a commercial edge to the agency's funding.

The agency itself has high expectations of its companies; among the ninety-seven HPSU firms it invested in last year, it expects more than 1,600 new jobs and more than €300 million in annual sales.

While large companies (those employing more than 250 people) made up only 0.2 percent of the enterprises in Ireland in 2009, they accounted for 31 percent of employment.[73] Both the state and its partner venture firms are playing the long game, aiming to create companies in that 'large' bracket.

Some in the industry have concerns about a state body having so much involvement in every level of venture funding. The sheer size of the agency's involvement means that its decisions – on which sectors to focus on or what development strategies to follow – have consequences for a massive number of early-stage companies. And its focus is on job creation, unlike VCs, which are focused on capital returns.

Scaling Problems

For all the positives of the Irish start-up ecosystem, there are problems with scale. Ireland simply does not have a big-enough market for many start-ups.

At the 2013 Web Summit, amidst all the noise, Oisin Hanrahan was keeping a low profile. Although he raised one of the biggest start-up funding rounds among Irish entrepreneurs in 2013, many people have not heard of him. The reason? His company was built in Boston, not Dublin.

Hanrahan is among a number of young, talented Irish entrepreneurs who have made a home across the Atlantic, preferring to build a company in the US rather than in Ireland. A graduate of Trinity College Dublin, Hanrahan is behind Handybook, an online platform which allows users to book pre-approved cleaners and handymen. In October 2013, the firm announced that it had raised $10 million in a new round of investment led by General Catalyst Partners and Highland Capital Partners.

Hanrahan, who originally went to the US to attend Harvard, said that while he has given plenty of thought as to whether he could have built Handybook from Ireland, ultimately he just does not believe it would have worked. His business is based upon having a large consumer market and, in truth, Ireland just wasn't big enough.

Given the difficulty involved in setting up a business, any small margin in a company's favour can make a big difference. For Hanrahan, being in the right location made a big difference. In the summer of 2014, as Hanrahan sought to move his company into the European market, it was London, rather than Dublin, that he chose.

Not that Hanrahan is downbeat about the Irish tech community. He sees a major opportunity for Ireland, as long as companies are focusing on the right types of problems.

Despite millions of euro in investment from venture capital firms, the government and state agencies such as Enterprise Ireland, there has not yet been a real breakout start-up success story. Two brothers from Limerick seem set to be that success, with a company valued at over $1 billion. There is however, a catch: they built their company in Silicon Valley. John and Patrick Collison are about as high-profile as it gets among Ireland's tech expats. Their company, Stripe, is one of the hottest start-up companies in the US.

Like Hanrahan, John also attributes his company's Stateside location to market size, saying Stripe is based in California rather than Dublin due to the size of the US market, rather than sources of funding. The San Francisco Bay Area alone, where Stripe is located, is home to more than 7 million people, compared to 4.5 million in Ireland. While John has been away from Ireland for years, he has been keeping an eye on the scene back home and described the difference between when he left and now as like the difference between night and day.

Where to, Next?

The Silicon Docks area still doesn't have a breakout star to really illustrate the potential of the Irish start-up sector. But there have been Irish successes. NDRC graduates Soundwave have become a hit on both sides of the Atlantic. Companies such as Logentries, Datahug, Intercom and Trustev may not be household names, but they have raised significant funding and gained a lot of traction. What there has not been is an exit or an initial public offering (IPO) that has hit the headlines.

Sean Blanchfield has already been through one successful start-up and is onto another. As a co-founder of gaming tech company Demonware, he has seen what a successful start-up exit looks like. Now Blanchfield is back in the game with PageFair, which helps companies solve the problem of ad blocking online, and is already finding plenty of interested partners and backers.

Blanchfield says capital gains tax is a major issue when it comes to attracting founders, as the high rate acts as a disincentive to investment. Then there is the state influence on the funding sector, which has created an awful lot of companies and funded them through venture funds using state money, creating a somewhat artificial funding cycle. It is also a simple reality that any start-up that wants scale needs to look outside of Ireland.

Blanchfield is hopeful that Ireland can foster a break-out start-up that turns into a billion-euro company, but for now at least, the jury remains out. The talent is there, so too the ideas and certainly the capital. For Ireland's budding start-up ecosystem, the odds are stacked in a manner that any entrepreneur would gratefully accept. That, however, is no guarantee of success.

8.

Global Gatherings of Tech Leaders: The Web Summit and F.ounders

Emmet Ryan

To the casual observer, the Web Summit is a success story that was certain to reach the top from day one. The truth is far closer to that of any other start-up story. There was a lot of hope, little money, plenty of gumption and a solid dose of failure.

Web Summit founder Paddy Cosgrave knows only too well the fate awaiting some of the businesses that exhibit at his event in the RDS each year. He's lived it. In 2007, his political media start-up MiCandidate caught plenty of attention and interest. The idea of a website with information on every candidate running for the 2007 general election in Ireland was almost cute, but the interest garnered showed it had real business potential.

A good idea that people like can get you far, but making it profitable is another matter entirely. Cosgrave went for a pan-European model ahead of the 2009 European elections, charging candidates to publish their information. It didn't catch on and, in October of that year, Cosgrave exited the business.

It was from what Cosgrave learnt along his journey with MiCandidate that the Dublin Web Summit was born. He had learned that running a start-up can be an isolating experience. YouTube was his only resource to hear great talks, and the only way to meet international investors, journalists and other tech entrepreneurs was to fork out several thousand euro to attend a conference abroad.[74]

Along the way, he had also made connections. He met other people with business ideas, some whom had already achieved or gone on to big things. When you can lean on the founders of Spotify, Wordpress and YouTube for advice, you've got a pretty solid contacts book. But Cosgrave didn't want their advice. He wanted to get their bodies to Dublin.

Prior to exiting MiCandidate, Cosgrave had persuaded potential clients from big European companies to come to Dublin for a look-see. They were still coming over, and he had to arrange something for them. So he organised an event in the Berkeley Court Hotel, and even bagged Minister for Communications Eamon Ryan to speak at it. The event was named the Dublin Web Summit.[75]

Over the following eight months, Cosgrave organised two similar events, with Craigslist founder Craig Newmark, Wordpress founder Matt Mullenweg and Bebo founder Michael Birch among the speakers. While they also used the name Web Summit, their place in the canon is different to the larger-scale showcases that would follow.

Like many of the firms that display at the Web Summit, Cosgrave started out running the events business from his bedroom, in a house he shared with five others. One of his housemates was David Kelly. Originally from Dunmore East in County Waterford, Kelly was working in finance in 2010, when his employer went bust.

Cosgrave asked Kelly his help in organising the tech conference, and Kelly was happy to have something to do for a few weeks after the rough end to his previous gig. For him it was a simple, short-term brief. He got to work with his mate for the guts of a month or maybe two and pad his CV. A long commitment couldn't have been further from his mind. While Cosgrave and, to a lesser extent, Daire Hickey hold substantial profiles in Irish life, Kelly is the quiet man of the triumvirate of Web Summit founders.

The First Major Event

The autumn 2010 Web Summit would be larger scale than the first three had been, and it was this event that put the technology conference on the map. It was set for October, 600 people were expected, and Chartered Accountants House was booked as the venue. It was at this time that the third and final member of the core team was brought on board, with the aim of bringing the event to a global audience.

Working as a freelance journalist, Daire Hickey wasn't easy to persuade. He had been one of the first freelancers at *The Journal.ie* and was getting work with several newspapers. The future was looking reasonably promising for him. But Cosgrave's persistence eventually convinced Hickey to work two days a week on what was, at the time, a run-of-the-mill conference in terms of scale – albeit, with a better calibre of speaker.

Cosgrave wanted Hickey to commit his time to guaranteeing interest from the media in Ireland. That took Hickey a couple of days. Job done. So what next? The decision to pursue international interest was natural, but what it delivered was far beyond the group's ambitions.

Bloomberg TV quickly got on board, buoyed by the guarantee of an interview with Twitter co-founder Jack Dorsey. Along with Bloomberg, CNN and *Wired* also attended the conference, meaning it was pretty much set on both the business and tech coverage fronts.

Setting up the event was a friends-and-family operation, with Kelly, Hickey, and Cosgrave's other halves, siblings and parents doing everything from erecting lighting to ensuring there was ice for speakers' glasses. Less than thirty people spoke at the event and all of the attendees came from one country – Ireland. There was only one workshop and one evening networking event.[76]

Among the speakers at the event was YouTube co-founder and chief executive Chad Hurley, who took to the stage to advise tech entrepreneurs to surround themselves with great people. It was at the event that Hurley announced his resignation as CEO of the video-sharing site, which took the Web Summit organisers by surprise.

With some of the world's top news organisations on hand to report it, the resignation became an international news story, and brought global media coverage to the Web Summit.

It was also in 2010 that the F.ounders event premiered, and cemented itself as a must-attend conference – if you get an invite, that is.[77] That same year, Bloomberg coined the 'Davos for Geeks' nickname of F.ounders.

The Wi-Fi broke, the audiovisual system didn't work, there was no food and there were very few overseas investors, but the 2010 Dublin Web Summit and F.ounders events were still hailed successes. In fact, the *Daily Telegraph* quoted one social gaming entrepreneur as saying that, 'the only invitees missing were Princess Diana and God'. [78]

As a result, Cosgrave, Kelly and Hickey decided to see where the events could go from there. They did not expect that the 500-person event would increase in size by forty times in the four years that followed, with the likes of Elon Musk flying in from the US and paying to have his Tesla car flown over too.

The Events Scale New Heights

Since that kick-off event in 2010, the Web Summit has grown rapidly. More than 1,500 people attended in 2011, with Amazon chief technology officer Werner Vogels, LinkedIn co-founder Eric Ly and Mikael Hed, chief executive of Angry Birds maker Rovio, among the line-up of speakers.

The 2011 F.ounders included a trip to Áras an Uachtaráin for a meeting with President Mary McAleese, and an evening at the Guinness Storehouse where Riverdance composer Bill Whelan gave a speech, followed by an exclusive performance by the Irish dance troupe. Skype founder Niklas Zennstrom is said to have donated the sails of his Fastnet Race twice-winning yacht to be made into bags for conference goers, and Bono participated in a pub crawl.[79]

It was at that year's event that rideshare-and-taxi app Uber locked down a funding round. In the surroundings of the Shelbourne Hotel, venture capitalist Shervin Pishevar signed a deal with Uber CEO Travis Kalnick to invest $26.5 million in the start-up. Three years later, Uber is valued at $18 billion.

By 2012, momentum was picking up, with 3,500 attending the Web Summit and SmartThings making its first big splash. The start-up prize went to the home automation start-up. Two years later, SmartThings was snapped up by Samsung for $200 million.

Once more, F.ounders played its part in attracting the big guns. Jim Sheridan and The Edge brought the A-list-Irish-success-story factor, while former US Treasury secretary Larry Summers and Google's Wael Ghonim – whose social-media postings galvanised protest in the Egyptian revolution – beefed up the international contingent.

History was made at the 2013 Web Summit, with the Nasdaq

stock exchange opened from Ireland for the first time. The opening bell was rung on stage in the RDS by Taoiseach Enda Kenny. The event attracted more than 10,000 attendees, including the CEOs and founders of some 200 start-ups, which had collectively raised a total of $5.9 billion in venture capital funding.

The headline moment for the 2013 Web Summit came from Trustev, the Cork-based anti-fraud firm, which announced a seed-funding round of $3 million – one of the largest for any European start-up in 2013. With 335 jobs announced at the event, not far off the attendance at the first Web Summit, the event was a clear win for the stakeholders who had got on board early.

Much as Uber's presence in 2011 was a sign of big things to come, so too was the attendance of another start-up – smart-thermostat maker Nest Labs at the 2013 Web Summit. Several months later, the company founded by Tony Fadell was acquired by Google for $3 billion.

Dropbox founder Drew Houston, AOL chief executive Tim Armstrong, Stripe founder Patrick Collison, Kaspersky Lab founder Eugene Kaspersky and Cisco chief technology officer Padmasree Warrior were among the speakers in 2013, with billionaire technology entrepreneur Elon Musk closing out the two-day event.

Musk, the co-founder of PayPal, Tesla and SpaceX, drove Taoiseach Enda Kenny into the RDS's main hall in a Tesla electric car. Their fireside chat there was reported widely, with the Taoiseach asking Musk for advice on how to steer Ireland, and Musk promptly telling him to ditch third-level fees for certain courses.

The 2014 event attracted 22,000 attendees from 109 countries, with speakers including actress Eva Longoria, supermodel Lily Cole, Paypal co-founder Peter Thiel, U2 frontman Bono and former Apple CEO John Sculley.

Managing Growth

While the Web Summit itself has grown in raw numbers, F.ounders hasn't shifted from its focus on keeping a fairly limited core together, with numbers rising since the start but not on anywhere near the scale of the Web Summit. For three days, approximately 220 of the world's top tech founders come together in one place. Everyone stays in the same hotel, there are pub crawls, dinners, and, because it's not just some wild frat party, there are talks too. It's all about creating ways for participants to meet each other, whether during a talk or over dessert.

At its heart, F.ounders is the same kind of concept as the Web Summit, just a whole lot smaller. The goal is for those who made the trek to Dublin to go away thinking it was good and that they made some business connections. Cosgrave says F.ounders is all about quality. It's about getting the right people in the room, in a relaxed environment. They know that everyone they are meeting is relevant to them from a business point of view.

In contrast to F.ounders, the Web Summit has seen phenomenal growth in attendance, which has created challenges. There are now hundreds of speakers each year, almost a thousand start-ups and thousands of attendees. No one involved foresaw the event growing to this scale this fast. Having started with three staff, the Web Summit now employs more than a hundred on a full-time basis.

The aim of the organisers is to make the Web Summit and F.ounders the best events of their kinds in the world. That can be a challenge, since every attendee has different goals. Negative feedback tends to ring louder than positive. Events tend to attract the same commentary as restaurants. If you have a good meal, you tell two people; if you have a bad one, you tell a dozen people or, worse, you tweet about it and it reaches even more. If you're a founder of a major tech firm, that tweet can reach hundreds of thousands of eyeballs within seconds.

That's the pressure facing the organisers of the Web Summit and F.ounders. Everything has to be the best because the organisers, their events and their reputations are only as good as their last conference. Hickey, Cosgrave and Kelly have to keep treating each event like it's their first, despite the increasing scale. By their own admission, in hindsight, they would have done some things very differently with some of their past events. For this organisation is about constantly learning and focusing on doing things better.

The Web Summit's lanyards are custom-made to be shorter than normal, so people maintain a better eye level. If it's higher up the chest, it's closer to where you're naturally looking. They pay this kind of attention to everything on the ID: from the font choice to the text size to the line spacing. Making the Web Summit a success every year requires the organisers to get right down to that level of detail.

The Night Summit

Where a lot of conferences go wrong is that they think their duty of care to attendees lasts only from 9 AM to 5 PM and, after that, it's 'Good luck, we'll see you tomorrow.' Anyone who has ever been to a conference, however, knows that even after the business part is over, there are still conversations you want to have.

Some of the best networking is done in hotel lobbies, at dinner and over drinks. The boardroom formality isn't as important as seeing what the person you are working with is like in real life. And if people are coming from all over the world to attend the Web Summit, they will want to see more of Dublin than just the RDS.

Since its inception, the Web Summit has worked to make the

night-time scene part of the core of the event. The displays are in the RDS; the night-time events are in the city. Every year, the event takes over numerous pubs across Dublin's city centre.

It was initially a purely organic part of the event but, with the organisers' penchant for tweaking, the emphasis on bringing pubs on board has become more focused. Exhibitors and attendees are getting more involved on this front, with the likes of Spotify and Microsoft holding their own Night Summit events in the city centre.

Even how this entertainment is delivered is carefully planned. It's common across the world, but particularly in the United States, for large tech conferences to finish with a big-name band playing. The likes of Imagine Dragons and Green Day can command hundreds of thousands of dollars to play the closing night of a conference. For event organisers, the idea may seem logical: attendees shell out plenty for a ticket, so give them something big as their final memory of the event.

The Web Summit, however, has deliberately avoided going down this route and not because of the exorbitant fees charged by the acts involved. Every step of the day is designed to enable better networking. A gig can be fun, but talking to people is hard if there's a large crowd and a big-name act. Live music is used at some of the Web Summit's pub events, but always on a smaller scale, to ensure that if people want to talk it is easy.

Changing Roles

As with any start-up, the brains behind the Web Summit brought different skills to the table and came in at different stages in their lives. Cosgrave had exited MiCandidate, Kelly had just suffered an unwelcome end to his work in the finance sector, and Hickey was barely up and running in journalism. Cosgrave is the obsessive frontman, Kelly the more laid-back macro-thinker, and Hickey is the glue guy, the one who brings both sides of the table together while also having good rapport with the media.

Their roles have changed substantially since the project began. All three had to be able to do some of everything when resources were limited. As the events have grown – and so too the number of bodies readily at their disposal – each member of the core team has been able to focus more on his primary strength.

Cosgrave still looks at every email the company sends out to woo attendees and sponsors, but he isn't carrying projectors or ice buckets. He is, by his own admission, obsessed with tweaking and improving things. This constant sense of dissatisfaction is what leads to aesthetic tweaks, such as the custom lanyards. Attention to is at the core of Cosgrave's role.

Right from the point F.ounders or Web Summit attendees land in Dublin, Cosgrave wants to manage their experience. Dublin's relatively small size means the organisers can count on security staff in the airport mentioning the event when visitors land but, once attendees get out of the arrivals lounge, Cosgrave wants to manage every move they make – up to and beyond slipping lanyards around their necks.

Cosgrave still makes sure taxi drivers are briefed before the event. Since taxi drivers are the first people in Dublin that most visitors are likely to have substantial conversations with, the Web Summit organisers see getting them onside as an opportunity that's not to be missed. Cosgrave describes taxi drivers as nearly the most important PR people in the country. When you come off your flight, they are honest brokers, the ones who will say the nice things but also fill you in on everything that's gone wrong in Ireland lately. That's a point of contact he still wants to influence. Detail remains everything for him.

In contrast, Kelly is focused on the broader picture for the Web Summit. Working with sponsors directly and looking at overall event-organisation are his key areas. Kelly doesn't want the public persona of Cosgrave or Hickey. His focus, instead, is on working with the sales team.

While Kelly still has a part to play on the ground during the conference itself, he's not involved in the event-management side of the business any more. He doesn't hustle for business internationally; his work is almost exclusively confined to the Web Summit's Dublin headquarters.

Hickey's life has taken a more dramatic turn. The Cork man is now based in New York, a move which was part of the Web Summit's strategic expansion, with the city seen as a key target market for advertising and consumer-focused tech start-ups. With further expansion on the cards following the firm's first event in Las Vegas, the Web Summit team see having one of their key men permanently in the US as critical to their organisation's growth.

For all the travels and movement in his life, the success of the start-up scene in his own home city of Cork is still important to Hickey. He takes particular pride in the trainload of start-ups that travelled up from the Rebel County to the RDS for the 2013 Web Summit.

The Changing Nature of Communication

The timing of the Web Summit's success wasn't all about serendipity. Technology and social media have played key roles. Events like the Web Summit can now reach wider audiences than they would have been able to ten or twenty years ago.

Most of you reading this have substantially more computing power in your pockets than was used to send astronauts to the moon. Keep in mind that the same sort of progress has been made in communications architecture, and you'll see that conferences have a new arsenal at their disposal.

Growth on a global scale has been made easier through the networking opportunities presented by such advancements in technology. It may sound basic, but Cosgrave simply wouldn't have had the tools he needed to connect with people in the same way had he been working with 1980s technology.

Conferences still existed, but their growth was limited by the communications tools available. The opportunity presented by advances in communications technology, often referred to as 'serendipitous' by Hickey, is a reflection of the time. It's also at the heart of why the Web Summit matters in the evolution of the Silicon Docks concept.

The one fundamental connection between all of the businesses that have sprung up in that part of Dublin is their focus on communication. From AirBnB to Twitter, AdRoll to Facebook, the growing market in this part of the city is around firms with communication at the heart of what they do. This environment has involved multi-tier coordination among industry and government bodies, but it has also seen events, like the Web Summit, develop.

Looking Abroad

From its beginnings in Dublin, the Web Summit has grown to become a globetrotting series of events.

Having already run tech conferences in New York, Berlin and London, the organisation ran its first event in Las Vegas in May 2014. The team opted to go 'off-Strip': instead of choosing one of the glitzy hotels along the Strip, which are home to the biggest tech shows in the world, they decided to work out of the city's downtown, which Hickey described as a lot like Dublin, with a growing start-up culture.

Next on the list for the Web Summit team is Asia, with an event planned for Hong Kong in 2015. All of this growth will require bringing more talent on board, be it back in Ranelagh, in New York, or at another site. The Web Summit's effort to get the best employees can't stop because, if it does, the competition will be waiting.

9.
Spill-over Effects

Philip Connolly

Take a walk down Barrow Street in the south Dublin docklands, and you will find two very different Irelands.

At one end of the street is Boland's Mill, a long-idle derelict building that was the scene of fighting during the 1916 Rising, when it was occupied by republicans led by Eamon de Valera. Taken over by the National Asset Management Agency (Nama) in 2012, it is a grey monument to the past.

At the other end of the street, several glass-fronted buildings connected by a skybridge form the home of Google's European headquarters. Here, talk of recession and gloom has been replaced by innovation and the energy of youth.

The Dublin docklands remains a place of contrasts. That, however, is changing.

Sitting just opposite the studio where U2 recorded 'Sunday Bloody Sunday', Ireland's 'bad bank' is trying its hand playing property developer on a scale it's never attempted before in Dublin. The plan is to turn Boland's Mill and the surrounding docklands into a version of London's Canary Wharf business district, at a cost of about €1.5 billion.

For years, the state's toxic debt agency has chased property developers to the ends of the earth, taking control of their properties and building up a vast portfolio of real estate.

Among those assets was a series of massive land banks and

unfinished projects around the Dublin docklands. It is a stretch of land that has previously scuppered the ambitions of a host of big-name developers. But things have changed. As tenants such as Facebook and Google have moved into the area, it has been transformed.

In the summer of 2014, Nama announced that it was prepared to spend €1 billion of its own cash developing residential and commercial properties in Dublin's docklands. For Brendan McDonagh, Nama's chief executive, having spent years tracking down assets and managing them, here was an opportunity to build something in an area that stacked up financially.

The docklands area was recently granted the status of a strategic development zone (SDZ) by An Bord Pleanála, a move that eased the way for large-scale development by making it easier to obtain planning permission. Within the docklands zone there are sites which make up over twenty-two hectares of development potential – enough space for 3.5 million square feet of office space and 2,600 apartments – and Nama owns or is linked through loans to 70 percent of those twenty-two hectares.

Property Plays

Ask anyone in the property game and they will tell it straight: office space in the docklands is the most sought after and expensive in Dublin.

That is why Nama and a host of large-scale property firms are so enamoured with the area. If they can capitalise on the demand, the lots dotted around the docklands that for so long seemed destined to be nothing more than a reminder of past follies could prove to be golden geese.

The docklands has a chequered development history, and

many of the sites host half-built developments and a complex ownership structures. Property titans like Liam Carroll, Johnny Ronan and Harry Crosbie all had ambitious plans for the area, many of which were undone by Ireland's economic calamity.

But if anyone needed evidence of the potential of the area, they need only look at the Dublin commercial property market over the past few years. After years of stagnation, the area is booming.

While many tech giants have already settled in the area, demand from new tech players and existing firms looking to expand is feeding the market. In the summer of 2014, Facebook moved from its original Dublin office into a much larger space at 4 Grand Canal Square, just beside the Bord Gáis Energy Theatre, to accommodate its growing workforce.

According to commercial property consultants CBRE, the volume of take-up achieved in Dublin in the first quarter of 2014 was the highest in a decade. In just three months, a total of sixty office lettings were signed in the capital. Principally, this was down to the popularity of the docklands.

For Nama and the developers and property funds within the docklands, the demand-fuelled rise in price has proved to be a vital fillip in a fluctuating market. The meteoric rise in prices in the area, however, means it is becoming too popular and expensive for many firms.

In late 2012, e-reader firm Kobo was fast outgrowing its Dublin office space, and gave itself two months to find new digs. The company's needs were simple and typical of many growing tech firms: an open space close to the city centre with plenty of light, which wouldn't require too much time or money to fix up.

Seven months and more than thirty property viewings later, Ian Black, director of research and development at Kobo, which has its European software development centre in Dublin, admitted that the initial two-month deadline was a little naive. What Black found was that within his price range, the offices were usually either very old or very unloved.

The firm wasn't alone. Black asked around and found that other tech companies were finding the same problem.

If you want to go out to East Point or Sandyford, two business parks on the fringes of Dublin, then you will find more than enough space. But if you want to be in the city centre, never mind the docklands, things get quite difficult. In the two years since 2012, the market has only gotten more competitive.

Tech firms such as Kobo may be among the price hike's victims, but they are also its principal cause. The primary driver behind the sharp resurgence in Dublin's office market is its booming technology sector.

Tenants leased 602,679 square feet of office space in the first three months of 2014 – a 21 percent increase over the same period in 2013 according to research by commercial real estate services firm Jones Lang LaSalle. Technology companies led the charge, taking up 41 percent of that space, the vast majority of it in the Silicon Docks area.

Earlier this year, rents for prime or Grade A offices, the type that are dotted around the docklands, spiked to more than €42 per square foot, a 31 percent increase in just over a year. As companies like Kobo expand and new firms set up shop, property of prime quality in the sought-after area is getting much harder to come by.

Even as rents continue to rise and leasing accelerates, construction remains at a relative standstill. So supply is dropping, with agents admitting that when a company comes in with specific requirements in a set geography, they really struggle. As a result, being a neighbour of either Google or Facebook, is proving to be very expensive.

When people look at the overall vacancy rate, which remains well in excess of 10 percent, they think there's lots of space, according to agents. The reality, however, is completely different. While there may be millions of square feet of office space still available in the city, most of it wasn't well maintained through the downturn and would require significant work to meet the standards sought by the largely US-based technology firms driving the current uptick in demand. For a growing tech firm, it just isn't worth the effort to do that work.

For companies looking to set up operations in Dublin or expand out of their existing spaces, the simplest answer is to look further afield than the docklands and its surrounding areas. That however, has its own problems.

Location can mean more than just convenience for technology companies. Demand is high for top-quality software engineers and programmers. The suburbs can be a tough sell with this crowd. In an effort to attract top talent, modern tech firms offer slick offices with pool tables, foosball games, and kitchens stocked with espresso machines and snacks. But they also need attractive locations.

People Traffic

More than anything, the opulent and playful offices of the docklands that are the preserve of the tech elite are a sign of change. Over the past decade, Ireland has lived through two parallel stories: a fall from grace and the rise of a new outward-looking technology sector.

US tech giants such as Google, Facebook and Twitter have backed Ireland as a perfect gateway into Europe. It would be naive to ignore the tax incentives as a major factor in their choice, but they also always cite another critical factor – the availability of talent.

While the number of people on the live register shows the influx of multinationals has not solved the jobs crisis, the availability of work in some of these iconic tech brands has acted as a beacon of hope for graduates and experienced professionals. For every difficult announcement the government makes about cuts and austerity, the IDA counters with news of a company expanding into Ireland or increasing its workforce.

Google and Facebook together employ more than 3,000 people. Both firms have attracted international talent in a wide range of functions, from sales and user operations to finance and recruitment. Those two firms, however, don't tell the full story of the rise in employment in the docklands.

The influx of big multinationals has become self-perpetuating, with each new company having a halo effect, attracting others, big and small. The industry has begun to feed itself, as the influx of companies has brought a great change in the culture of start-ups to Dublin.

Tech firms look for people of a certain profile, people who are entrepreneurial and can adapt quickly. With Facebook and Google working as a perfect finishing school, that type of talent is both attracted from abroad and fostered. This has led to the IDA targeting emerging firms, such as Dropbox, which don't yet have the profitability to avail of the tax breaks but do need the talent.

It isn't quite Silicon Valley, where tech workers earn almost three times as much as their Irish counterparts, but Silicon Docks is witnessing its own war for talent. It can now be very challenging for multinational technology companies with lofty plans for dizzying expansions to secure what they term 'top-tier' talent.

According to recruitment experts, tech firms retain a reputation as places where employees are treated with care, granted unprecedented working freedom, and given plenty of opportunity for advancement, in Ireland and overseas. The flip side of this is an expectation that employees will learn and perform.

Beyond the personal aspects of the job, tech firms also have a reputation for paying very generous salaries. This, however, has had an unforeseen consequence for start-ups.

A good programmer can expect to earn a salary significantly in excess of €60,000 in a tech multinational, while an indigenous start-up would be doing well to offer much more than half that.

For both offices and talent, it seems, the competition in the docklands is fierce. Not that it is just the tech firms that are hoovering up office space and talent. Other firms have arrived to the docklands too.

The legal profession seems to like the surroundings of the south docks, with a lucrative new client base on the doorstep. Just up the road from Google on Barrow Street, business 'legal eagles' Mason, Hayes and Curran have moved in. Further down on the riverside is top corporate law firm Matheson. Also along Sir John Rogerson's Quay are the law practices McCann FitzGerald and Beauchamps.

Global financial firms have also started moving to the area. State Street has set up in one of the more striking offices at the end of docks, while HSBC has taken up residence in One Grand Canal Square next to the Bord Gáis Energy Theatre. Consultancy firm Accenture has its offices in the same building overlooking the Grand Canal basin.

Over the past few years, the docklands have also played host to a variety of more niche businesses, from a number of firms specialising in high-frequency trading to a couple of now-wound-up football funds holding international professional players' economic rights.

Taking Up Residence

The influx of people into tech jobs in the docklands has also had effects on what was a faltering residential market, as the upward swing in the property market is not just confined to office space.

Having paid around €50 million in 2011 to buy and develop the site beside the Bord Gáis Energy Theatre on which The Marker now sits, Brehon Capital made its money back on the deal without even selling the hotel. The adjoining residential development, which was included in the original deal, was sold for in excess of €50 million, more than €10 million over the original asking price. The deal for the block of eighty-four luxury apartments a stone's throw of Facebook's new offices involved a fierce bidding war, as domestic and international funds sought exposure to one of the more popular areas to live in the city.

The Marker Residences is one of a number of recently built apartment complexes in the area which have proved increasingly popular among both buyers and renters. Built almost fourteen years ago by Liam Carroll's Zoe Developments, penthouses atop the Millennium Tower – Dublin's tallest residential building – are estimated at a value of upwards of €1 million. In 2013, a two-floor unit at the top of the building, which stands over the Grand Canal basin, was sold for €802,000. Since then, property values have continued to rise. On the other side of the basin, apartments at Hanover Dock are being sold for more than €400,000, while at the Alto Vetro tower rents can top €3,000 per month.

Not that the rise in property values is just confined to the docklands itself. In Ringsend, an area near Grand Canal Dock that went into serious decline as the shipping trade dried up, prices are also on the rise. In 2012, you could have picked up a two-bedroom house for approximately €200,000, but by 2014 you had pay almost €300,000, according to the Irish property price register.

Changing Face

At the bottom of Lower Grand Canal Street, a short stroll from Google's multi-hued office complex, sits 3FE. A decade ago, the coffee shop would have seemed out of place on the street, which is occupied by council flats and convenience shops. Now the place is packed every day with a heady mix of people whiling away the hours on top-end laptops, and young, affluent coffee aficionados.

Set up by Colm Harmon in 2011, 3FE is the brainchild of the former trustee officer for professional investment funds in Dublin's financial district, who realised he was in the wrong business. The high-end coffee shop is just one example of how the tech set has changed the face of the docklands.

By the end of 2008, Grand Canal Dock had seen an influx of cafes and restaurants, and was fast developing a name as a dining hub. Dieter Bergmann led the way, with the opening of Riva Restaurant on Hanover Quay that same year. Real Gourmet Burger opened on Gallery Quay, followed by Asian restaurant The Crystal Boat. These joined Ely, Herbstreet, Milanos, Bridge Bar and Grill, Il Valentino, KC Peaches and Café Java.

Some of these did not survive the recession. By 2011, however, the docklands were rising out of the embers of the property crash. When Prince Albert II of Monaco visited Grand Canal Dock with his fiancée, Charlene Wittstock, a theatre and a raft of new restaurants and cafes had opened in the area.

Sitting underneath the railways arches near Grand Canal Dock, just outside the glow of the of the Bord Gáis Energy Theatre's lights, is Pizza e Porchetta. The restaurant has recently added another flavour to the local palate with its brand of high-quality pizzas and Italian food. Beside the theatre is The Marker, Dublin's newest high-end hotel. When the hotel finally opened, it drew to a close a development saga typical of the Celtic Tiger. Built during the boom by Pierse Construction, no operator could be secured for the hotel, and it sat idle until 2013. Now it plays host to tech entrepreneurs, bankers and venture capitalists, as well as tourists.

In front of the hotel is also the incongruous sight of people windsurfing between office buildings and apartment blocks. Availing of the cattle hold of the *Naomh Eanna*, a retired ferry more used to travelling the sea between the mainland and the Aran Islands, is water-sports company Surfdock.

Surfdock started out in 1992, transforming the Grand Canal basin from an aging relic into a water-sports playground. Originally a small windsurfing school, the firm has grown alongside the area, and it is now common to see people kayaking, stand-up paddleboarding, windsurfing and wakeboarding around the Dublin docklands.

Once a focal point for maritime businesses, where cargo barges began their journeys to the west of Ireland, the waters of the Grand Canal basin are a hive of activity again.

Spillover Effects on Wider Ireland

The influx of tech firms has changed Ireland in ways no one could have imagined.

Located above a Centra in Portlaoise is the office of Ireland's data-protection commissioner, a role that once mostly involved dealing with consumer complaints. Over the last few years however, Billy Hawkes, Ireland's long-serving data-protection commissioner, became one of the main men in European data regulation.

As the digital industry has grown, personal data has become a more valuable commodity. Who exactly owns a person's data, and what exactly they can do with it, have become massively contentious points across Europe. Given that most of the firms involved in the debate have their European headquarters in Dublin, the city has become a focal point

Change is afoot: the EU is moving towards a more centralised regulatory structure. Yet, for now, as torrents of information are collected about the majority of the population of Europe on a daily basis, the Irish regulator remains at the centre.

In the summer of 2014, the EU's top data-protection official, Isabelle Falque-Pierrotin, said things had reached a point where US technology giants such as Google and Facebook were too big to be left in the hands of individual national-privacy watchdogs. Essentially, digital companies are now beyond the capacity of any one national regulator.[80]

Many of the world's top digital companies now fall under the remit of Helen Dixon, who replaced Hawkes as Ireland's data-protection commissioner, when he retired in September 2014.

Prior to his retirement, Hawkes said the stricter data-protection regime that is currently working its way through Brussels could have profound effects on how companies are regulated. Progress, however, has been slow. The EU's current rules were drafted nearly two decades ago to deal with an industry that has, since then, completely transformed.

To date, EU nations have dragged their heels over revamping the law. In the meantime, case law involving digital companies is evolving as the Irish courts have also become involved in the debate.

Change for the Better

An area once known for being the centre of a vibrant shipping trade, by the 1990s the docklands had become a relic of a time long gone by. The Dublin Docklands Development Authority was tasked with redeveloping the site in 1997, but the game really changed in 2004 when Google showed up in town.

Facebook followed shortly afterward and began an influx of firms into the area, bringing thousands of young employees, many of them from continental Europe, along with them. It is a long way from the area's former life as a working-class maritime hub, but the docklands, it seems, has been irrevocably changed by the tech firms that have found new homes around it.

The change for the most part has been for the better, and the same can be said for wider Ireland. Data-protection might be an issue, property prices might by high, and the war for talent is ongoing. But the arrival of Silicon Valley giants has given new life to the docklands and a sense of hope to an economy that has been through turmoil.

10. Silicon Republic: The Broader Context

J.J. Worrall

Find the yellow surfboard then take the stairs to the basement. It sounds like the directions given to an unwitting criminal by undercover police officers in *Hawaii Five-O*. However, if you're on West 44th Street between 8th Avenue and 9th Avenue in New York City then it will lead you to Réunion.

The best – and, let's face it, quite possibly the only – surf bar in Hell's Kitchen, it could easily be missed by a research-light tourist. Thankfully, though, many visitors to the city who've found themselves thirsty, hungry or simply curious and in the locality have been directed there after checking out the raft of positive Réunion reviews on their Yelp app.

Having co-founded the urban guide/business review/essential travel companion a little over a decade ago, Jeremy Stoppelman announced in June 2014 that Yelp was to be the latest major Web company to set up offices in Dublin, creating in excess of a hundred jobs in the process.[81] He described the opening of the Dublin offices – which will act as the fulcrum of the company's European operations – as a 'landmark moment' for Yelp, one which came during a trade visit to Silicon Valley by Taoiseach Enda Kenny.

Located on Hatch Street in the south of the city, Yelp will be based a mile or so away from Silicon Docks. Spread the net further around the city and into the suburbs and you'll find many other technology and online companies that chose not to locate in the docklands. PayPal, eBay, Yahoo!, Dell, Microsoft, Oracle, Adobe, IBM and SAP have settled in Blanchardstown, Cherrywood, Sandyford, Citywest, East Wall or elsewhere in the capital. Many too have decided to locate additional offices in the north, south and west of the country, as well as in Dublin's neighbouring counties.

A prime example of this country-wide strategy is Hewlett Packard (HP). Arriving in Ireland in 1995, HP's president and managing director, Franz Nawratil, initially had modest employment targets, peaking at a few hundred employees. Skip forward nineteen years, and HP now has offices and R&D units employing 4,500 people around the island of Ireland, in Dublin, Kildare, Belfast and Galway.

Back in Silicon Valley, the day before Kenny and Stoppelman smiled for cameras in mid-2014, the Taoiseach was busy welcoming an announcement from HP that it was to bump up its Galway workforce in Ballybrit Business Park by another hundred, taking the number of employees at the company's European software centre there to 650.

By no means comparable with the east coast in terms of technology-related employment, the west of Ireland was, however, home to one of the first communications multinationals to find a home on Irish soil. It's also where we begin our breakdown of the Silicon Republic.

The West

The Irish government might prefer it to be remembered as the year the nation joined the European Economic Community, but recollections of 1973 often begin, instead, with a helicopter journey.

That the journey began set off from Mountjoy Prison in Dublin and the passengers included a pilot being held against his will and three escaping IRA volunteers should explain its place among newsreel highlights of the year.

That same year was, however, also when a company that would become one of the cornerstones of technology employment in Ireland for the guts of the next four decades arrived on these shores.

Northern Electric's move to Galway in 1973 was followed by continued expansions which saw the telecommunications company greatly swell its employment numbers over the next twenty-five years. By the time the dot-com bubble was ready to pop in the late 1990s, the Canadian company – which had been renamed Nortel – employed over 800 people in its west-of-Ireland facility, which acted as its European manufacturing base.

At its peak, the company reported annual revenues of $30 billion and had worldwide staff numbers exceeding 90,000. In Galway, Nortel was a fixture in the local community, even helping to sponsor the region's annual arts festival for almost twenty years. In June 1998, Nortel acquired Bay Networks – which had offices further south, in Shannon, County Clare – for $9.1 billion, bringing a further ninety Irish employees onto its payroll.

It was around that time, however, that rumblings of major trouble within the company started to make global headlines. In September of that year, it was announced that 3,500 employees would be let go. Irish headlines emphasised that those working in Galway would not be affected, as manufacturing operations there were on 'a separate track' to the company's other European businesses.

What followed over the next decade is perhaps a lesson worth remembering for other multinational employees should they hear such reassurances from employers in the future.

Soon after the September announcement, initial steps to

reduce Nortel's numbers in Galway saw the introduction of a voluntary redundancy scheme. With a large number of 'low-tech' manufacturing roles within the company moving to Cwmcarn in Wales, this became a viable escape route for many experienced employees.

Over the next twenty-four months, Nortel's total Irish workforce fell rapidly. The company had its Galway and Shannon operations, a relatively small sales office in Dublin as well as its facilities north of the border in Antrim, where over 2,000 people were employed. Hundreds were forced to leave its Galway operations, in particular. By mid-2002, employee numbers in that corner of the company's Irish business were down to about 300.

On 14 January 2009, the news that many expected finally came through. Nortel was no more. A simple message on the company's website stated that it had 'initiated creditor-protection proceedings in multiple jurisdictions'. The move came a day before a $107 million interest payment was due.[82]

Nortel – a company with roots stretching back to 1895 – was sold to a communications-technology business less than a decade in existence that December, at a price of $900 million. Already employing 200 people in Sandyford, south Dublin, Avaya took over responsibility for 300 former Nortel workers in Galway and about 500 in Belfast and, thankfully, continues to thrive in both locations.

The Arrival of Cisco

In November 2006, Cisco announced its intention to hire 200 technical and engineering staff over the following thirty-six months in Galway. Initially, fifty staff were added to an operation that was to be integral to its future worldwide R&D activities, with a heavy focus on the company's data, voice and video products.[83]

Described by Minister for Enterprise Micheál Martin as a 'superb win for Ireland', Cisco's arrival in Galway was to involve the company spending €30 million on an R&D facility. Former Nortel employee Barry O'Sullivan, a Cork native who would later appear as an investor on the Irish TV version of *Dragons' Den*, headed up the move for Cisco.

Based in Silicon Valley, O'Sullivan ran Cisco's $2-billion unified communications division and decided to locate a key part of the operation in Galway. He cited the proximity of third-level institutions like the National University of Ireland Galway, the Digital Enterprise Research Institute and the University of Limerick as key to the decision.

The following April, reports surfaced of the company poaching Nortel staff for the facility, with over a dozen employees having apparently made the switch.[84]

In recent years, Cisco announced an expansion of the Galway R&D centre, with a further €26 million being spent on its facilities in Oranmore, creating another 115 jobs. All told, the company currently employs approximately 300 people between its offices there and in Eastpoint Business Park in North Dublin.

Alongside HP and Cisco, some of the other big international names which have gone west over the past decade or so include gaming company Electronic Arts and one of the business world's other major software-solutions innovators, SAP, which announced plans to enter the west of Ireland on April Fool's Day 2003, having arrived in Dublin on the same date six years earlier. The company now employs more than 1,400 people in four offices in Galway and the capital.

On the whole, though, the west hasn't been overrun by tech success stories. The return of Philip Martin to his native Leitrim in 1999, however, is certainly worth mentioning. Founder of Cora

Systems, which specialises in project-management software used by local authorities in Ireland and Britain, Martin initially hired three people. Now he has ten times that number working from the company's Carrick-on-Shannon base.

The South

'Do you actually believe that every home is, within the short term . . . the next few years . . . going to have its own personal computer?'

His arrival in Ireland wasn't quite the showstopper that John DeLorean's arrival in Belfast the year before had been. However, when Steven Jobs – as he was then called – touched down in Cork in late 1980 to open Apple's manufacturing facility in the city, local media rushed to the scene in droves.

As work continued in the assembly plant behind the Apple co-founder, Irish TV-and-radio stalwart Pat Kenny asked Jobs the big question about whether computers would make it into the average home. 'We base our theory,' said Jobs, 'on the fact that we make personal computers that can be used irrespective of the location.' Media coverage included tales of mini computers and housewives throwing away cookery books to take recipes from flickering PC screens.[85]

While there are no news reports to indicate that a *Fahrenheit 451*-esque cookery-book purge followed in Cork and the surrounding counties, the new factory in Hollyhill in the north-west of the city brought the Californian company to Ireland's attention for the first time. The facility was a IR£7-million gamble intended to one day create 700 jobs, though only nineteen employees started on the first day in the 44,000-square-foot plant.[86]

The first manufacturing hub for Apple outside the US, the facility initially caused anxiety among union members because it lacked a time clock to 'monitor the coming and going of the workers' as *Irish Independent* reporter Dick Cross put it. Apple, it seemed, wanted to create a more relaxed atmosphere.[87]

By the end of its first decade in Ireland, the company had over a thousand employees here, and the facilities at Hollyhill had expanded rapidly to over 340,000 square feet. In the 1990s, as Apple struggled to maintain its place at the front of the PC market, some production responsibilities would move to Singapore and across the Irish Sea to Wales.

In 1997, the company signed a strategic partnership with Microsoft, a move that today might seem utterly alien, but at the time was seen as a massive boon for ailing Apple. Two years later, Apple management at Hollyhill confirm that they would be letting go of a huge number of employees.

What happened next, though, is a story anyone with even a passing interest in technology, music, phones or tablets will be well aware of. As Apple transformed its business in the early years of the next decade – with iPods then iPhones then iPads causing people to queue overnight – Hollyhill too became a different animal.

What was once a manufacturing base to rival anything else in Europe became the company's European centre for support services. By the eve of the Apple's release of the iPad in mid-2010, the company had broken the 2,000-employee barrier there. Plans for a further 300 Hollyhill recruits – brought in to help production of the latest Mac Pro initially – were also announced around the same time.

As with most of its announcements about its Irish base, Apple didn't seek press attention for the creation of these roles. This approach to publicity sets it apart from the majority of large tech players in Ireland. Come right up to early 2014, with employee numbers standing at 4,000 in the Cork area, Apple still remains a relatively quiet giant in our midst.

Perhaps this has something to do with the touchy subject of tax, and just how much of it Apple pays to the Irish state and to the United States. When Tim Cook – successor to the now sadly departed Jobs – arrived in Hollyhill in January 2014 to meet Taoiseach Enda Kenny, they both instantly faced questions on the subject.

This was, perhaps, to be expected after a US Senate committee found in 2013 that Apple had paid a corporate tax rate of 2 percent or less here.[88] At the end of September 2014, the European Commission said it believed Ireland had conferred a tax advantage on Apple. The Commission said two tax arrangements agreed between Ireland and Apple amounted to illegal state aid.[89]

EMC Opts for Cork

It shouldn't be forgotten that another one of Ireland's biggest employers also arrived in Cork in the 1980s. EMC opened its facilities in Ovens in 1988, less than a decade after it had been founded in the US.

Built on the idea of creating data-storage solutions, they began with a sixty-four-kilobyte memory board. By the time the company arrived in Cork, EMC had begun to specialise in slightly more advanced storage options, and engineers there were at work on the product that would become EMC's flagship for much of the next twenty years: the Symmetrix storage array.

The year the Ovens facility opened its doors was also the year EMC's eventual global CEO and chairman, Michael C Ruettgers, arrived at the company. A great believer in the Cork facility right up until his retirement in 2005, Ruettgers, it's fair to say, wasn't so sure the company was meeting expectations when he first arrived.

Ruettgers famously gave an assembled group of workers

airline-style sick bags. Referring to complaints EMC had been getting about faulty memory disks, Ruettgers held his bag aloft and, so the legend has it, said: 'The quality of our products makes me want to puke'.[90]

At the time, EMC was manufacturing products for IBM and Unisys mainframes and could not afford any more bad press. Thankfully, though, a swift upswing in the company's fortunes saw its staff in Ovens swell to 120 a little more than four years the company's arrival there.

The Boston-headquartered business saw its revenue increase by 331 percent in 1993. By July 1994, it was estimated that the company had pumped $40 million into its manufacturing base in Cork.

Staff numbers grew and the company had 650 employees on its Cork payroll by mid-1997, a period which saw EMC announce its decision to double its manufacturing capacity worldwide. It was a move which would result in the Ovens base growing two-fold, coming in at a whopping 400,000 square feet, enough space to accommodate 500 extra employees over the following eighteen months.

By this point, EMC Ireland was selling into all global markets except the US. Worldwide, the company had 5,500 employees and sales topping $8.8 billion by the turn of the century.[91]

It wasn't long before the facilities in Ovens expanded to almost 600,000 square feet, seeing investment by the company in Ireland pass the $400-million mark. A massive $3 billion in software and services was generated from Cork in 2000. However, plans to increase the number of employees there to 1,800 over the next few years failed to materialise, with the company cutting 1,100 jobs globally and numbers in Ovens hovering around 1,300.

In 2004, EMC stepped in to acquire virtualisation market leaders VMware for $625 million. VMware then announced plans to create 350 technical-support and back-office jobs in Ballincollig, Cork, in 2007.

A partnership with University College Cork to create the EMC Research Europe centre followed, and the company built a new €100-million office in Mahon, County Cork. Including offices in Shannon, Dublin and Belfast, the company now employs 3,200 people here.

Turbulent Times for Dell

While EMC's journey in Munster has been relatively steady, Dell has had a somewhat more turbulent time in the province. When Dell arrived in Ireland in October 1990, to set up a manufacturing facility in the Raheen Industrial Estate in Limerick, the PC maker had to take second billing in most news reports on investment that day.[92]

While Michael Dell's decision to invest IR£7 million in Limerick was trumpeted by Minister for Industry and Commerce Des O'Malley as a great boost to the area, it was dwarfed in the headlines by the arrival of the Hualon Corporation of Taiwan. That company was going to invest IR£47 million in a new textile-manufacturing base, also in Limerick.

Dell was taking a chance on refurbishing a facility that had lain vacant since Atari had left the area six years previously, a move that had left 270 people unemployed. Dell estimated it would be employing 300 people in Raheen within three years.

The company, however, grew at an astonishing rate. The manufacturing centre employed almost 900 people by 1996, with a further 370 telesales staff located in the east of the country in Bray, County Wicklow. That same year, Dell PCs worth IR£500,000 were hijacked by a Dublin gang.

It was also in 1996 that a IR£20-million expansion of the Raheen facility was announced, along with the promise of 500 further full-time jobs. Sales from the manufacturing base rose 35 percent year on year, and Michael Dell said the company's Irish operations 'excelled in every measure of performance'.[93]

During the next two years, sixteen production lines for PCs, notebooks and other devices powered the Limerick facility beyond the 2,000-employee mark. A near-$90 million development was announced in 1999, which would bring a further 1,700 new jobs to Raheen, pushing Dell's Irish workforce over 5,800 when taking in Bray and offices in Cherrywood Business Park in South Dublin.

However, these enormous employment figures weren't to last. In May 2001, 200 administration jobs were lost in Limerick. A further 125 job losses were shared across Raheen, Bray and Cherrywood soon after. A year later, the company sought 150 voluntary redundancies in Limerick.

In November 2004 the company would consolidate its Bray facility with Cherrywood, creating a marketing-and-sales hub for Europe. Incremental job losses continued and, by early 2007, there were fears that a thousand Limerick-based manufacturing positions could be at risk. The rising cost of doing business in Ireland was one of the factors said to be involved.

With 4,500 still employed by the company in April 2008, it was then announced that 250 more employees would be let go between Dublin and Limerick. As Christmas 2008 closed in and the global financial crisis took hold it became increasingly obvious that a huge number of jobs were to be lost in Limerick.

In early January, it was announced that 1,900 people would be let go in Raheen. This was despite the facility making an operating profit €11.5 million over the previous twelve months. It was a move that had ramifications for the livelihoods of thousands of other workers beyond Dell's gates in Raheen and the story dominated news programmes on national TV for weeks. In all, 1,100 employees would remain in Limerick to coordinate manufacturing, supply-chain activities and other functions for Europe, the Middle East and Africa.

It's worth remembering, though, that as of autumn 2014, Dell's Irish operations employ more than 2,500 people between Raheen, Cherrywood and Cork (where Dell bought up Quest Software in 2012), in the company's global services centre to software, finance, sales and marketing operations and a cloud-engineering centre.

For all the ups and downs, Ireland is still – as Dell's president for Europe, the Middle East and Africa, Aongus Hegarty, told the *Irish Times* recently – 'pivotal' to the company as a whole.[94]

More companies have flocked to the south of Ireland, including Huawei, the Chinese telecoms giant hoping to usurp Apple in the devices market (in the Far East at least). The company has been a network-hardware provider in the Irish market for twenty years. It became more widely known as 'the dongle company' and finally a multi-device-making 'upstart' with annual turnover not too shy of $40 billion and more than 150,000 employees worldwide.

Founded in 1987, Huawei actually landed in Ireland more than a decade ago – in May 2004 – and employs seventy-five people across two offices in Cork and one in Dublin. Much of the work here is focused on research and development. Huawei reports that it's put more than €4.57 million into R&D activities in Ireland in 2014.

While Huawei has found itself on the wrong end of some cyber-espionage stories, one of the other international names to call Cork home now is a business which is regularly at the centre of unmasking cybercriminal activity.

Trend Micro, founded in 1988 in the US, came to Cork in 2003. From an initial seven workers, the IT-security company has grown and employs almost 300 today. Research labs are at the forefront of the company's work, helping to bring down cybercriminal enterprises from across the globe from its offices on the Model Farm Road.

The East

'What is IBM?' It was the question that headed up several newspaper adverts for the company in Ireland in 1962. Well, at that point, it was a company which was six years into its time in Ireland and in great need of men. That was their preference – it *was* 1962. They had to be aged between twenty-one and thirty, with at least a finals standard City & Guilds in telecommunications in their back pocket.

The good people at International Business Machines Ireland guaranteed that starting salaries would not fall below IR£650 too. Having developed its first computer in 1944 – used to calculate naval gun trajectories – the company arrived in Ireland just as it released the 350 RAMAC Disk Storage Unit, IBM's first commercial hard disk drive. It has remained an influential presence in the Irish business marketplace ever since.[95]

The changing face of IBM over the decades would be reflected in its Irish operations. There was the 1980s PC boom, with manufacturing of the 5150 and other machines in Dublin. The 1990s saw workers here help with the R&D that would, along with more everyday functions, allow IBM to create the Deep Blue supercomputer that got the better of reigning world champion Garry Kasparov in a chess game.

At present, IBM employs more than 3,000 people across a range of sites around Ireland. The imprint the company has here is huge. Sales, services, manufacturing, marketing and R&D are all carried out in the company's Irish outposts.

By the time Bill Gates announced to a Paris audience that his business was going to set up international production facilities in Dublin, it's not entirely clear how many people would have been able to convincingly answer the question 'What is Microsoft?'

It was 10 May 1985, and the company's operating system was on board IBM computers being sold here. Microsoft's plans for Dublin were initially conservative. Less than twenty employees were expected to be hired when its first office opened here the following year.[96]

The move came at a wonderfully opportune time as, over the course of the next eighteen months, Microsoft Windows version 1.0 hit the market, while, in the boardroom of the company's Washington headquarters, Microsoft was preparing to go ahead with its initial public offering.

By March 1988, the company had eighty-six people working in its production facilities in Sandyford, south Dublin, and it was announcing a IR£1.5-million expansion that would see its numbers swell beyond 110 as the 1990s rolled around.

One of the founding companies of Sandyford Business Centre, by late 1991, Microsoft had almost 500 people working for it there. Fast forward to 2014 and Microsoft has – subject to certain local authority approvals – been granted permission to create what some have termed a 'mega-headquarters' in nearby Leopardstown for its 2,000-strong workforce.

A mix of full-time employees and contract staff, they currently work in four buildings in the surrounding area. Microsoft Ireland performs a range of functions, including software-development, localisation, sales, marketing and finance for Europe, the Middle East and Africa.

Long-time supporters of Ireland such as Microsoft, IBM, EMC, Apple, HP and Dell, having largely stayed put during the country's good and bad times, have been a saving grace for plenty of Irish governments through the years. Announcements such as the Microsoft mega-HQ make them appear rooted here, and, during the most recent financial crisis, these companies' occasional jobs announcements and milestone celebrations provided some of the few sources of positive news for the government.

Intel on the Rise

Take, for example, Taoiseach Brian Cowen's car journey to Leixlip, County Kildare, on 27 November 2009. In the midst of rising national debt and banking catastrophes that savaged Ireland in the wake of the financial crash, Cowen was in town to help Intel celebrate its twentieth anniversary in the country.

Since it had landed in Ireland in 1989, each time Intel had announced another 10,000-square-foot extension to its campus over the next two decades – or had broken news of 100, 200, 500 or more jobs arriving in Kildare – you could be certain there was someone high up (quite often the very highest up) in the Irish government front and centre, smiling in the press pictures and making a gag about being able to spell hard disks or big data but not being overly aware of what they entailed.

There's no doubt that Intel's impact in the eastern region of the country over the past two decades has been seismic. That time has seen the US microprocessor maker turn a 360-acre stud farm into a sprawling campus employing in excess of 4,500 people and investing billions of euro in the area.

Back in Dublin, while a jaunt around the capital's business parks wouldn't make any tourist itinerary, there's plenty to admire in the employment many other companies have created in the city.

Familiar names will include gaming giant Blizzard in Eastpoint

Business Park, as well as their neighbours there like Cisco, Citrix, Ergo and Oracle, not to mention Yahoo!, which were formerly the offices of Overture before Yahoo! acquired the search start-up in July 2003. Speak to people who were in the IDA in the early part of the last decade, and they'll tell you that Overture was head-hunted at the same time, and with just as much vigour, as the centrepiece of Silicon Docks: Google. Both officially arrived in early 2003.

Then, in the west of Dublin, there are the twin behemoths eBay Europe Services and PayPal Europe Services, both of which are based in Blanchardstown. Between them, they employ more than 1,500 people for customer, corporate and administration services. The two companies also have large operations in County Louth. In 2014, PayPal announced that it would create 400 new jobs in Dundalk, in addition to the 1,000 positions it had announced in 2012. This will bring the combined PayPal and eBay employee numbers in Dundalk to 1,850 by 2018, and the overall employee numbers in Ireland to more than 2,900.

The North

Leaving the six counties paying tax to Britain aside and concentrating on the Silicon Republic in its strictest terms, Cavan, Monaghan and Donegal in the north reflect the reality of many other counties which many will argue have often been underserved in terms of Internet infrastructure and have consequently been overlooked by potential investors.

Go to Wicklow, Wexford, Louth or Longford, and the predominant industries range from food, fuel, investment firms and insulation, to businesses based on waste, water and timber.

Head for Tipperary, Kerry, Waterford and Clare and, in amongst some familiar tech names, start-up hubs and technology hardware manufacturing bases, the focus of revenue in the area is divided between the hospitality sector, pharmaceuticals, steel fabrication, agriculture and food.

Waste management, foodstuffs, medical devices and textiles are some of the big employers in Sligo and Roscommon in the west, while Manorhamilton in Leitrim hosts Mirror Controls International, which arrived in 2001 and is a leading global manufacturer of mirror technology for the automotive industry, employing about 170 people.

In the north it's a similar story. Cavan has the insurance giant Liberty, having once been the centre of Sean Quinn's empire before it went belly up in the crash. Meat, animal foodstuffs and plastics are the other big employers in the county.

In Monaghan, meanwhile, Monaghan Mushrooms is a huge employer in the county, with around 2,750 people working in offices there, as well as in facilities across the border in Armagh. Monaghan Veterinary Laboratory employs just under a hundred people, having begun business in 1966, while companies focused on agribusiness and transport logistics also provide employment in the area. There's also Combilift, a company based on the combination of a forklift and a sideloader, which employs approximately 130 in the area.[97]

In Donegal, one of the standout businesses is Pramerica Systems Ireland. The software-development company concentrates on areas like consultancy, training, project management, analysis, design, development and support. A subsidiary of US-based financial services group Prudential Financial, which had more than $1.1 trillion of assets under management as of March 2014, it first set up in Donegal in June 2000 and now employs more than 1,100 people in the area.

It can only be hoped that future investors – maybe even

Stoppelman's emerging Silicon Valley neighbours – scouting for locations in Ireland will also see beyond Dublin. Perhaps then a more balanced Silicon Republic can emerge, one not so invested in Dublin and Cork.

11. The Future of Silicon Docks

Ciara O'Brien

Just a few short years ago, Ireland was in the grip of an economic crisis. Rising debt levels, a property-market crash and the problems that had emerged in the country's banks were stifling the country's economy.

The boom had turned to bust in a matter of months, and young people were emigrating in their thousands. Unemployment soared as job losses were announced with alarming regularity, and thousands of businesses went to the wall.

Office buildings, newly completed, were uncertain of ever attracting the hoped-for tenants, and the shells of half-finished buildings littered Dublin's city centre. Now things are looking up again, and that is partly due to the tech sector.

Take a stroll around the area that has become known as Silicon Docks and you'll see a vibrant business sector. It's approximately one square mile in size, but its influence is far more significant.

On Barrow Street, where Google is based, property values are soaring as the US tech giant sprawls over several buildings. In its shadow are companies such as Airbnb, Facebook and Twitter, along with incubators such as Dogpatch Labs that have made Silicon Docks their home. These stand shoulder to shoulder with start-ups and accelerators that have increased in number over the past couple of years. This is the result of some hard campaigning and marketing of a country that, until recently, was a tough sell.

The IDA set up an emerging business unit in 2010, under its Horizon 2020 plan, to identify and bring to Ireland early-stage, fast-growth companies to Ireland. It wasn't the easiest time to persuade firms that Ireland was a good location. In the grip of the Troika bailout, the country's image was tarnished and the economy was rocky. Nevertheless, the IDA team set about finding companies that could provide Ireland with the next wave of tech firms, and help the country's economy grow. The plan of attack included networking, marketing events and one-to-one meetings with firms. Although some of the firms were not yet making big money, there was a good chance they would be in the future, and when they did, the IDA wanted to make sure they were established in Ireland.

Back in 2010, the senior vice president of IDA Ireland's emerging business division, Barry O'Dowd, was looking at the docklands as an area of opportunity. The vital thing, he thought, was to get to a stage where the agency had a rolodex of companies that it could use to attract others, and to get a few key names to use as a starter pack. That came with the likes of Indeed.com, HubSpot, Etsy and others. While the jobs announcements have been small – a hundred jobs here, twenty-five jobs there – over time they have begun to add up to a more impressive overall figure. The emerging business unit has notched up more than 2,000 jobs to date and, by the end of summer 2014, it was about to close its hundredth project – a major milestone.

Although they're still looking to hit the jackpot with the next Google or Facebook, the signs are encouraging with companies like Zendesk, which announced the opening of Irish office in 2012 and a data centre last year, raising $100 million in an IPO in May 2014. It is companies like this that will be the future cornerstones of Silicon Docks, an area that has gathered together some of the biggest names in the tech industry.

But Silicon Docks isn't limited by a postcode. It's more than that, Barry O'Dowd says. He describes it as a great piece of branding, which has become much bigger than the docks itself, rather than a geography. The tech companies are following the water out from a core area in the docks, down the River Liffey and the Grand Canal. As Silicon Docks expands, Dublin's reputation as a tech hub is growing too.

Facebook has already outgrown its original Dublin base, as, it seems, has Twitter. In June 2014, Mark Zuckerberg's social network moved its Dublin office from Hanover Quay to Grand Canal Square, doubling the potential capacity of its Irish base to a thousand and making it the biggest operation outside of its headquarters in California. Twitter, which set up on Pearse Street, could also be on the move, as it seeks growth of its own. LinkedIn is another company planning expansion in Dublin. The professional networking site acquired a 17,507-square-metre site at Wilton Place in late 2014, where it intends to build a new international headquarters. Squarespace has already moved once and is looking to move again, while Riot Games has moved into larger spaces three times.

There has been some progress in creating more office space in the area. In May 2014, Dublin City Council gave the go-ahead for a major development of the docklands area that will not only include thousands of square metres of office space, but also more than 2,500 homes. Nama is to finance a €150-million redevelopment of Boland's Mill, which will include a fourteen-storey office block. The agency has also submitted a planning application to develop 42,500-square-metres of offices and apartments at Hanover Quay and Sir John Rogerson's Quay. With a fast-track planning process in place and some of the property in the area already in public ownership through Nama, things could get going quite quickly.

The Tech Scene in Ireland

Now that three waves of three different types of tech firms have landed on Irish shores, the question is what the fourth wave might be.

The first wave began with IBM, which first opened an office here in 1956 and now employs about 3,000 people in Ireland.

Before the iPhone was even a footnote on Apple's balance sheet, the US tech firm was setting up its Irish operation in Cork, opening its doors in 1980. Intel, meanwhile, made its first move into Ireland in 1989 and in early 2014 announced it had invested some $12.5 billion in its Irish operation to date.

The second wave of companies coming to Ireland were involved in the software side of things, with firms such as Oracle and Microsoft bringing thousands of jobs with them. And the third wave, the most recent additions, are the born-on-the-Internet companies – like Facebook, Google, Etsy and Dropbox – that are currently building their staff counts in Ireland and growing every year.

There were a few casualties along the way. Gateway closed its manufacturing plant in Dublin in 2001. Dell moved its manufacturing operations from Limerick to Poland in 2009, although it has built up employment in Ireland in the provision of services. At the last count, Dell employed more than 2,500 people in Limerick, Dublin and Cork.

Part of the newer wave of companies, games firm PopCap Games closed its Dublin studio in 2012. Although the job losses were lower – around a hundred jobs went with the closure – it was more about the signal that it sent. Since a major effort was underway to grow the games industry in Ireland and establish it as a hub for the sector, the news that PopCap Dublin was no more was a blow. But, on the whole, things for the Irish tech sector have been reasonably bright.

There are a few possibilities for a fourth wave of tech firms. IDA Ireland has been busy trying to predict what will catch on next, with data analytics among the top contenders. It's an area Ireland could excel in and, in Barry O'Dowd's view, it could be a real winner for the country. Analytics brings together maths, behavioural sciences and communication skills. Already, there are third-level courses in analytics, and government investments in big-data research centres, such as the Technology Centre in Data Analytics, a collaboration involving DIT, UCD and UCC.

Analytics is closely interlinked with another another sector that might provide a fourth wave: the 'Internet of Things'. Smartphones have given rise to smart homes, where appliances, lighting and heating can all be controlled via the Internet. Wearable technology has increased in popularity, giving users instant feedback on their activity levels and heart rates, uploading it all to the cloud so it can be tracked across multiple devices. Cities are becoming smarter, as sensors manage traffic flow to reduce gridlock and air pollution, creating large swathes of data that need to be analysed.

Companies such as Cisco, which has a research and development centre in Galway, and Intel, which set up its own Internet of things division in 2013, have already recognised the value of the sector. According to Cisco, it could be a $19-trillion market, so even capturing a small corner of it would be a success. And to give it a bit of real-world application, Intel announced plans to make Dublin one of the world's most densely sensored cities, and to create an accompanying research platform making the data gathered by the project available to city residents and other interested parties. IBM has also been actively involved in researching smart cities, with a lab located in Dublin since 2011 that collaborates with top universities, cities and industry partners.

Looking at smaller firms and start-ups, the IDA was in advanced talks to bring smart-home firm Nest Labs here, but the company was bought by Google before a deal could be completed – yet another indicator that 'the Internet of Things' is going to be the next big market for the tech firms. But, with the sector very much in its infancy, there will no doubt be more companies for the IDA to target.

The Issue of Tax

Ireland's tax system has long been considered a major pull for firms, with a 12.5-percent corporation tax that is well below that of other countries. Germany, for example, imposes a rate of almost 30 percent, while the United Kingdom taxes companies at 21 percent, although effective tax rates can be lower once tax credits are taken into account. A recent study published by PwC found that Luxembourg had the lowest effective tax rate in the EU, at 4.2 percent.

Ireland's tax system has been a source of tension for the government on the political stage. Everyone from former French president Nicholas Sarkozy to US president Barack Obama to German chancellor Angela Merkel has had a pop at our low tax rates for corporations. Companies that relocate to Ireland for tax purposes are gaming the system, Obama said. The low tax rate helped fuel the banking crisis, according to Angela Merkel.

Whatever way you look at it, it's clear that the international community sees Ireland's corporate tax rate as something of an unfair advantage, and one they are keen to get rid of. When we were in the grip of the bailout, there was talk of a higher tax rate in return for concessions to get us out of our dire financial situation. It's an advantage that we've been exploiting for many years.

Long before the current debate over corporate tax rates, Ireland was using a low corporate tax rate to attract new investment.

But that's not the only thing that is attracting firms to Irish shores. Even before it became a hot political topic, almost every tech executive would, when asked, play down the importance of tax rates, preferring instead to discuss the benefits of an educated workforce, an availability of skills and that catch-all, 'a business-friendly environment'. That doesn't stop business lobby groups such as IBEC and other interested parties from urging the government to keep the competitive tax rate.

Tax experts say the pressure to increase Ireland's corporate rate has eased off for now. But, even if it had not, the thought of a small increase doesn't send them into a panic. That's because they also subscribe to the notion that Ireland's tax rate is not the sole reason companies come here, although it's a nice bonus. Also on the scale – along with the tax rate, educated workforce, availability of skills and business-friendly environment – is the fact that Ireland is an English-speaking economy and can act as a gateway for US companies into Europe.

A tax hike could be more of a danger to large foreign direct investments rather than start-ups, experts say. It could be argued, though, that the hope is the start-ups will eventually be the ones to attract FDI. So, while it may not be an issue for them now, it could become one further down the road.

PwC's John Murphy says the 12.5-percent rate is part of a suite of incentives that also includes Ireland's R&D tax credit and tax deductions relating to intellectual property. The tax credit and the ability to monetise and obtain a cash tax refund of the R&D tax credit might play a more important role in attracting start-ups to a certain extent, particularly when taxable profits are low. This is because start-ups can eliminate or reduce their social security contributions by crediting them against their R&D spend, and thus creating immediate cash flow benefits.

As for the attraction of start-ups to Ireland, Murphy says it's worth noting that some start-ups that are from jurisdictions where tax rates are comparable or lower still move to Ireland, lending some weight perhaps to the argument that it's not all about the corporate tax rate, although that rate does play an important part. Serbia's Nordeus is one such company, and is viewed as one of the IDA's success stories in attracting growing tech firms to Ireland. Serbia's current corporation tax rate stands at 15 percent.

Although the 12.5-percent rate may be considered safe for now, international corporate tax rules may be about to undergo a change prompted by an OECD review. In September 2014, the Organisation for Economic Co-operation and Development (OECD) set out plans to eliminate tax avoidance by big multinationals, by clamping down on the practice of shifting of profits to low- or no-tax jurisdictions. The OECD proposals aim to ensure that corporate profits are taxed where economic activities generating the profits are performed and where value is created. While the OECD proposals are not binding on governments, plans are afoot to develop a binding multilateral convention.

Following pressure from the OECD, along with the European Commission and countries such as Germany, France and Britain, the Irish government announced the closure of the controversial 'double-Irish' corporate tax loophole in October 2014. Multinational companies such as Google exploited the loophole to shift millions in profits offshore virtually tax-free. The government said all new companies registering in Ireland from January 2015 would also have to be tax-resident in Ireland. The mechanism would be phased out by the end of 2020 for companies already using it.

But tax experts also point out that while Ireland's regime is transparent – what you see is what you get, more or less – there are other ways to reduce the tax burden if companies are so inclined. What could change for Ireland as a result of changes to the global tax system is further transparency. Reputation is an important factor for the government, and that has taken somewhat of a battering amid talks of secret tax deals.

The closing off of the 'double-Irish' tax loophole was seen as a major move on that front.

Murphy says the move was welcome, helping the government to deal with some of the negative press attached to Ireland as an investment location, while also providing certainty to companies already based here on the remaining lifespan of structures, to allow them to plan for the future development.

Whether it will affect Ireland's ability to attract new investment has yet to be seen. In public at least, companies using the scheme are taking a 'don't panic' approach. The impact of the changes will be offset in part at least by new tax breaks and initiatives aimed at bolstering investment.

Murphy says from an 'Ireland Inc' perspective, the OECD proposals on profit-shifting should benefit Ireland in the long-run because they are focused on ensuring that value-creation, substance and tax outcomes are aligned. This, he says, should ensure that competitive tax locations like Ireland, and Dublin, in particular, remain at the top of the list of locations for investment and encourage more jobs and activity here.

Despite optimism that raising the corporate tax rate from 12.5 percent would have a limited impact on Ireland's ability to attract start-ups and tech firms, the only sure way to know is to raise the rate and watch the impact. And given the government's commitment to keeping the rate at 12.5 percent, it doesn't look like that will be a real risk any time soon.

The Next Billion-dollar Start-up

With so many start-ups in Ireland, could the next billion-dollar tech company come from these shores? There are plenty of Irish entrepreneurs trying to make that happen. However, the size of the Irish market and the opportunities available here mean that, realistically, budding billionaires must look further afield to grow their companies.

When it comes to business, it's a numbers game, and the hard truth is that the Irish market doesn't have what most companies need to build a billion-dollar company. With a population of around 4.5 million, it's smaller than most large cities in the US or even the UK. The US market is frequently seen as a must for Irish companies seeking to grow.

So which Irish start-ups have the potential of expanding beyond Ireland – grabbing the UK and US markets – and becoming billion-dollar businesses?

Soundwave is one of a new generation of companies built around the app economy. Founded by Brendan O'Driscoll, Aidan Sliney and Craig Watson, the music-discovery app tracks what people are listening to in real time. It has certainly grabbed attention. Soundwave now has offices in Dublin and San Francisco, and has been adding new features to its product and bringing it to new platforms such as Google's Android Wear.

Software firm Intercom, meanwhile, has also turned its attention to the US market. Co-founder Eoghan McCabe has been open about the company's ambition to become a billion-dollar company. No matter how optimistic you are about the Irish market, you have to admit it's a feat that wouldn't be possible simply relying on Ireland.

It's a situation that serial entrepreneur Pat Phelan is familiar with. At forty-eight years of age, the co-founder of anti-fraud firm

Trustev packed up and moved to New York to continue the company's expansion there. He was too old for the move, in his opinion, and should have done it at twenty-four, instead. But it was a necessary step for the business, and it is showing signs of paying off. Among Trustev's major new customers in the US is Radio Shack.

It's been a hard slog for the Trustev team, but so far at least it's been a satisfying one. Less than three years old, the company has already collected a number of awards, including the top prize in the start-up accelerator competition at SXSW in Austin, Texas in 2014. In 2013, the European Commission named the firm Europe's top technology start-up, and it has also had the honour of being pinpointed as one of *Forbes* magazine's hottest global start-ups. In short, Trustev is turning heads.

Ireland's Changing Start-up Scene

Building your own tech business has become one of those things people do; it's trendy, Phelan says. But building your own company is tough, with long hours and hard decisions required to even hope to make it successful. Phelan says that, more and more in Ireland, he sees some people getting caught up in the idea of the start-up and all that it brings with it, rather than focusing on the reality of building a company. The allure of being a start-up CEO has become more important than the allure of creating a start-up.

For some people, it's more about the personal brand rather than the company brand, about doing start-ups for the fame of it. It's a route that Phelan has actively avoided with Trustev. These days, he gives speaking engagements and other similar events a wide berth, saying that 'brand Pat Phelan' is dead, and the company has reaped the benefits.

Ireland's start-up scene has changed dramatically in the last few years, and not entirely for the better. Still, there are some positives, including improved supports and better funding opportunities than before the 2008 recession. Trustev has benefitted from some of the changes, including the emergence of accelerators such as Wayra, where it started off. But while incubators and other support programmes have become vital parts of Ireland's tech scene, Phelan says there is a trend for some start-ups to 'incubator hop' and never remove the training wheels. He puts it down to a fear of failure, a central part of the process. He says many start-ups will fail, and that's what people miss; they should accept it and use the failure to do something better next time. 'Fail better' is an often-used quote, but it seems that it's an appropriate one for entrepreneurs.

Selling Out Early

Compared to a just a few short years ago, it looks a bit sunnier for Irish entrepreneurs who are willing to put in the effort and commit to their start-up.

However, while Ireland may be a hotbed of innovation, the economy isn't reaping as much benefit as it could. That was an assertion made by head of the Irish Stock Exchange, Deirdre Somers, at the ISE's annual conference in 2013. Somers said there needs to be a rethink in Ireland's enterprise policy, if indigenous companies are to be supported. She wasn't talking about start-ups, but rather the mid-size firms, which are being snapped up by bigger firms from outside Ireland. She said this situation would continue, with Irish companies selling to international buyers, unless there is a rebalancing of incentives.

Somers called for a rethink on how Ireland taxes capital, share options and exits, and also new funding sources, as the problems in the banks impacted on potential for growth.

The Future

It's impossible to say for certain how Ireland's tech sector will look in ten or fifteen years. You'd be sticking your neck out if you were to confidently predict that the country would continue to attract high-value businesses and start-ups with massive potential. If we've learned anything in the past few years, it's that things can change in – almost – the blink of an eye.

While the tech sector didn't emerge unscathed from the economic downturn of the past few years, it's in better shape than many sectors that were considered parts of the bedrock of Ireland's economy in the Celtic Tiger years. With a vibrant start-up community and State agencies working to attract up-and-coming companies to Ireland, the foundations are being laid for the next wave of companies to grow and begin contributing in meaningful ways to the Irish economy. However, complacency is not an option.

Ireland is facing competition from other European countries, which are also building up their reputations as 'the next Silicon Valley', meaning we're constantly looking over our shoulder. It's not necessarily a bad thing; a little healthy competition keeps us on our toes. Above all, it's good for the start-up sector, as more supports and more recognition could lead to more growth. And it keeps the accolades rolling in: the easiest location in the EU to start a business, the most business-friendly tax regime in Europe and, at least according to *Forbes* magazine back in 2013, the best country in the world for doing business. It's a high standard to meet; let's hope Ireland can keep it up.

Notes

1. Slonaker, Larry. 25 September 1995. *San Jose Mercury News*. tbtf.com/siliconia1.html

2. Wieners, Brand and Hillner, Jennifer. 1998. 'Silicon envy'. *Wired*. archive.wired.com/wired/archive/6.09/silicon.html

3. Dublin Docklands Development Authority Community Liaison Committee. 'Report of 15 years of community gain obtained through the CLC influence on docklands development'.

4. Seanad Éireann. 27 February 1997. Debate on Dublin Docklands Development Authority Bill, 1996: Second Stage. debates.oireachtas.ie/seanad/1997/02/27/printall.asp

5. Forfás. March 2012. 'Evaluation of enterprise supports for start-ups and entrepreneurship'. www.forfas.ie/media/17042014-Enterprise_Evaluation_of_Start-Ups_and_Entrepreneurship-Publication.pdf

6. Enterprise Ireland. 'Ireland is a dynamic source of start-up funding'. www.enterprise-ireland.com/en/Start-a-Business-in-Ireland/Startups-from-Outside-Ireland/Why-Locate-in-Ireland-/Ireland-is-a-dynamic-source-of-start-up-funding.html

7. Ruane, Noel. 'Why Dublin is probably the best place in Europe to launch your start-up'. 2011. www.dhrp.ie/news/Why-Dublin-is-probably-the-best-place-in-Europe-to-launch-your-startup/#sthash.n63jv5kn.dpuf

8. Trinity College Dublin. 2014. 'Innovation: Trinity campus companies'. www.tcd.ie/innovation/exchange/campus-companies/

9. Inland Waterways Association of Ireland – Dublin Branch. 2010. '50th anniversary of the last trade boat, 1960 – journey to be recreated'. dublin.iwai.ie/news2010.html

10. Bunbury, Turtle. 2008. *Dublin Docklands – An Urban Voyage*. Montague Publications Group. www.turtlebunbury.com/published/published_books/docklands/grand_canal_docks/pub_books_docklands_gcd_gasometre.html

11. National Archives of Ireland. 'Dublin waters: the Liffey, the canals, and the port'. www.census.nationalarchives.ie/exhibition/dublin/waters.html

12. Moore, Niamh. 2008. *Dublin Docklands Reinvented: The Post-Industrial Regeneration of a European City Quarter*. Four Courts Press.

13. O'Carroll, Aileen. 2006. '"Every ship is a different factory": work, organisation, technology, community and change; the story of the Dublin docker'. *Saothar: The Irish Journal of Labour History*. www.academia.edu/226953/_Every_ship_is_a_different_factory_Work_Organisation_Technology_Community_and_Change_The_Story_of_the_Dublin_Docker_

14. GSP. 2014. 'Poised for redevelopment: Boland's Mill at Grand Dock in Dublin, Ireland'. www.globalsiteplans.com/environmental-design/infrastructure/bolands-mill-at-grand-dock-in-dublin-ireland-juxtaposition-of-old-and-new/

15. Moore. *Dublin Docklands Reinvented.*

16. Dublin Docklands Development Authority. 'Dublin docklands area master plan 2003'. www.dublindocklands.ie/files//business/docs/20070416100245_Part%202.pdf

17. Bunbury. *Dublin Docklands*. www.turtlebunbury.com/published/published_books/docklands/grand_canal_docks/pub_books_docklands_gcd_gasometre.html

18. Moore. *Dublin Docklands Reinvented.*

19. IFSC. 'About the IFSC'. www.ifsc.ie/page.aspx?idpage=6

20. Moore. *Dublin Docklands Reinvented.*

21. Irish Statute Book. 'Dublin Docklands Development Authority Act 1997'. www.irishstatutebook.ie/1997/en/act/pub/0007/

22. Dublin Docklands Development Authority Community Liaison Committee. 'Report of 15 years of community gain obtained through the CLC influence on docklands development'.

23. Dublin Docklands Development Authority. 'About us'. www.ddda.ie/index.jsp?p=99&n=138

24. Dublin Docklands Development Authority. 2000. 'Grand Canal Dock area planning scheme, 2000'. www.ddda.ie/index.jsp?n=172&p=124

25. Comptroller and Auditor General. 2012. 'Special report: Dublin Docklands Development Authority'. www.audgen.gov.ie/documents/vfmreports/77_DDDA.pdf

26. Moore. *Dublin Docklands Reinvented.*

27. Comptroller and Auditor General. 'Special report'.

28. Martha Schwartz Partners. 'Grand Canal Square, Dublin, Ireland'. www.marthaschwartz.com/projects/civic_institutional_dublin.php

29. Moore. *Dublin Docklands Reinvented.*

30. Dublin Docklands Development Authority Community Liaison Committee. 'Report of 15 years of community gain obtained through the CLC influence on docklands development.'

31. Comptroller and Auditor General. 'Special report'.

32. Dublin Docklands Development Authority. 'About us'.

33. Kelly, Olivia. 23 May 2014. 'Bord Pleanála approves fast track planning for Dublin docklands'. www.irishtimes.com/business/sectors/commercial-property/bord-pleanála-approves-fast-track-planning-for-dublin-docklands-1.1805715

34. Intel. 2009. 'Intel Ireland employees drive 20 years of excellence'. www.intel.ie/content/dam/www/public/emea/ie/en/documents/20-yr-booklet.pdf

35. Mac Sharry, Ray and Padraic A. White. 2000. *The Making of the Celtic Tiger: The Inside Story of Ireland's Boom Economy*. Mercier Press.

36. Telsis Consulting Group. 1982. 'A review of industrial policy'. The National Economic and Social Council. irishlabour.com/dublinopinion/Telesis_part1.pdf

37. Industrial Review Group. 1992. 'A time for change: industrial policy for the 1990s' (Culliton Report).

38. IDA Ireland. 'Interview with Sheryl Sandberg of Facebook'. www.youtube.com/watch?v=LBn41tQCIUg

39. Kennedy, John. 2008. 'The Friday Interview: Nelson Mattos, Google'. *Silicon Republic*. www.siliconrepublic.com/new-media/item/10836-the-friday-interview-nelso

40. Weintraub, Seth. 29 September 2010. 'Excite passed up buying Google for $750,000 in 1999'. *Fortune*. fortune.com/2010/09/29/excite-passed-up-buying-google-for-750000-in-1999/

41. Smyth, Jamie. 30 January 2003. 'Google close to deal on Dublin centre'. *Irish Times.*

42. Smyth, Jamie. 13 March 2003. 'Google to set up its European centre in Dublin'. *Irish Times.*

43. Smyth, Jamie. 7 October 2004. 'Founders preach Google gospel in Dublin'. *Irish Times.*

44. Keena, Colm. 12 December 2005. 'Google to announce 700 new jobs'. *Irish Times.*

45. 15 November 2006. 'Google announces major Dublin expansion.' *Irish Times.*

46. Weckler, Adrian. 28 September 2012. 'Google's Dublin workforce rises to 2,500'. *BusinessPost.ie.*

47. MerrionStreet.ie. 2013. 'Taoiseach officially opens foundry at Google'. www.merrionstreet.ie/en/Category-Index/Economy/Smart-Economy/taoiseach-officially-opens-the-foundry-at-google.51724.shortcut.html

48. Irish Times. 18 November 2005. 'Google saves millions in taxes thanks to its Irish operation'. *Irish Times.*

49. Mulligan, John. 6 October 2012. 'Google pays just €8m tax here by routing €9bn profits abroad'. *Irish Independent.*

50. Slattery, Laura. 25 July 2014. 'Google pays €27.7m tax at Irish subsidiary on €17bn revenue'. *Irish Times.*

51. O'Connell, Hugh. 9 January 2012. 'Good news: Limerick not wiped off the map'. *Daily Edge.* www.dailyedge.ie/good-news-limerick-not-wiped-off-the-map-323636-Jan2012/

52. Hickey, Shane. 30 May 2011. 'Ireland has a new Facebook friend: Zuckerberg turns tourist'. *Irish Independent*. www.independent.ie/irish-news/ireland-has-a-new-facebook-friend-zuckerberg-turns-tourist-26737690.html

53. McLysaght, Emer. 27 May 2011. 'Did you have pint with Mark Zuckerberg last night?' *TheJournal.ie*, www.dailyedge.ie/did-you-have-a-pint-with-mark-zuckerberg-last-night-144608-May2011/

54. McLysaght. 'Did you have pint with Mark Zuckerberg last night?'

55. Callanan, Neil. 20 January 2012. 'Facebook said to weigh doubling size of European headquarters in Dublin'. *Bloomberg*. www.bloomberg.com/news/2012-01-20/facebook-weighs-expanding-european-hq.html

56. Deegan, Gordon. 5 December 2013. 'Facebook staff reap €12m shares windfall – twice firm's Irish tax bill'. *Irish Independent*. www.independent.ie/business/technology/facebook-staff-reap-12m-shares-windfall-twice-firms-irish-tax-bill-29810986.html

57. Vardi, Nathan. 11 May 2012. 'The Facebook employee stock dump is on'. *Forbes*. www.forbes.com/sites/nathanvardi/2012/11/05/the-facebook-employee-stock-dump-is-on/

58. Kennedy, John. 29 September 2011. 'Facebook says "like" button is not being used to track people'. *Siliconrepublic.com*. www.siliconrepublic.com/new-media/item/23821-facebook-says-like-button

59. Europe-v-Facebook. 2012. 'UPDATE: Facebook breaches deadlines in Irish data protection report – users can now file a complaint at the European Commission Brussels'. www.europe-v-facebook.org/EN/Media/MELDUNG_EN_2.pdf

60. Kennedy, John. 10 September 2013. 'Ireland's data protection

commissioner works in eye of data storm. *Siliconrepublic.com* www.siliconrepublic.com/enterprise/item/34116-irelands-data-protection-c

61. Kennedy, John. 10 September 2013. 'Ireland's data protection commissioner works in eye of data storm'. *Siliconrepublic.com.* www.siliconrepublic.com/enterprise/item/34116-irelands-data-protection-c

62. Demos, Telis and Menn, Joseph. 28 January 2011. 'LinkedIn looks for boost with IPO'. *Financial Times.* www.ft.com/cms/s/0/59e47ba4-2a54-11e0-b906-00144feab49a.html

63.Collins, John. 30 September 2011. 'Ireland's answer to Silicon Valley'. *Irish Times.*

64. Mulligan, John. 18 March 2010. 'Irish base for US online gaming giant to boost our "smart economy"'. *Irish Independent.* www.independent.ie/business/irish/irish-base-for-us-online-gaming-giant-to-boost-our-smart-economy-26641913.html

65. Millar, Angela. 11 July 2011. 'Social games company Zynga launches in Dublin'. *BBC News.* www.bbc.co.uk/news/technology-13739404

66. McCaffrey, Una. 4 April 2014. 'Major global tech firms must now have an Irish presence, conference hears'. *Irish Times.* www.irishtimes.com/business/sectors/technology/major-global-tech-firms-must-now-have-an-irish-presence-conference-hears-1.1749303

67. Kennedy, John. 3 December 2012. 'Dropbox to locate international HQ in Dublin'. *Siliconrepublic.com.* www.siliconrepublic.com/careers/item/30526-dropbox-to-locate-internati

68. Kennedy, John. 5 March 2013. 'The online marketing model is broken, says HubSpot CEO Brian Halligan'. *Siliconrepublic.com*. www.siliconrepublic.com/new-media/item/31747-the-online-marketing-model

69. Kennedy. 'The online marketing model is broken, says HubSpot CEO Brian Halligan'.

70. Smith, Stephen. 16 May 2014. 'FDI – Ireland's 50-year overnight success story'. *IrishCentral.com*. www.irishcentral.com/business/technology/FDI—-Irelands-50-year-overnight-success-story.html#ixzz31tAf1l2t

71.Rusli, Evelyn M. 7 July 2011. 'The new start-ups at Sun Valley'. *New York Times*. dealbook.nytimes.com/2011/07/07/the-new-sun-valley-start-ups/

72. NRDC. 2014. 'NRDC named in top 2.5% of business incubators in the world'. www.ndrc.ie/news/ndrc-named-top-2-5-business-incubators-world/

73. Department of Finance. 'Economic assessment of the SME sector in Ireland'. www.finance.gov.ie/sites/default/files/12%2017%20Taxation%20of%20Small%20Business.pdf

74. Cosgrave, Paddy. 19 September 2013. 'A letter to Irish start-ups'. Web Summit blog. www.telegraph.co.uk/technology/technology-events/8097044/Tech-founders-gather-in-Dublin-to-plot-Eu ropes-future.html

75.Hunt, Joanne. 28 October 2011. 'Top scores for web event'. *Irish Times*. October 28 2011. www.irishtimes.com/business/sectors/technology/top-scores-for-web-event-1.633083

76. Cosgrave. 'A letter to Irish start-ups'.

77. Bryant, Martin. 24 October 2012. 'Is F.ounders still the Rolls Royce of technology events?' *The Next Web*. thenextweb.com/eu/2012/10/24/is-f-ounders-still-the-rolls-royce-of-technology-events/

78. Yiannopoulos, Milo. 29 October 2010. 'Tech founders gather in Dublin to plot Europe's future'. *Daily Telegraph*. www.telegraph.co.uk/technology/technology-events/8097044/Tech-founders-gather-in-Dublin-to-plot-Europes-future.html

79. Rowan, David. 2 March 2012. 'F.ounders – the conference that's "Davos for Geeks"'. *Wired*. www.wired.co.uk/news/archive/2012-03/06/founders-dublin

80. Bodoni, Stephanie. 10 May 2014. 'Google to Facebook seen as too big for EU nations' Privacy Czars'. www.bloomberg.com/news/2014-05-09/google-to-facebook-seen-as-too-big-for-eu-nations-privacy-czars.html

81. IDA Ireland. 6 June 2014. 'Yelp to create 100 new jobs in Dublin'. www.idaireland.com/news-media/press-releases/yelp-to-create-100-new-jo/index.xml

82. Reuters. 14 Jan 2009. 'Timeline: key dates in the history of Nortel'. *Reuters*. www.reuters.com/article/2009/01/15/us-nortel-timeline-sb-idUSTRE50D3N120090115

83. Cisco. 21 November 2006. 'Cisco to establish global R&D centre in Galway'. newsroom.cisco.com/dlls/2006/prod_112106.html

84. Collins, John. 6 April 2007. 'Cisco poaches Nortel staff for R&D unit'. *Irish Times*.

85. RTÉ (via EverySteveJobsVideo.com). 1980. 'Steve Jobs TV interview about Ireland's arrival in Ireland'. everystevejobsvideo.com/steve-jobs-tv-interview-about-apple-in-cork-ireland-1980/

86. Irish Apple Blog. 19 January 2012. 'When Steve Jobs and Apple first came to Ireland'. www.irishappleblog.com/when-steve-jobs-and-apple-first-came-to-ireland

87. Cross, Dick. 30 December 1980. 'Apple saga brings mini-computer to kitchen'. *Irish Independent.*

88. RTÉ News. 21 May 2013. 'Apple confirms 2% tax rate for two Irish subsidiaries'. www.rte.ie/news/business/2013/0521/451564-apple-tax-arrangements/

89. Irish Times. 30 September 2014. 'A bite out of Apple'. *Irish Times.* www.irishtimes.com/debate/editorial/a-bite-out-of-apple-1.1947923

90. Business Week. 14 March 1999. 'EMC: high-tech star'. *BusinessWeek.com.* www.businessweek.com/stories/1999-03-14/emc-high-tech-star

91. Krause, Reinhardt. 20 August 2014. 'EMC CEO Tucci building future; VMware spinoff doubted'. news.investors.com/technology/082014-714043-ubs-doubts-emc-will-spin-off-vmware.htm

92. Quinlan, Arthur. 23 October 1990. '1,000 jobs announced for Limerick'. *Irish Times.*

93. Murdoch, Bill. 11 November 1996. 'Electronics firms create 1,400 jobs'. *Irish Times.*

94. O'Brien, Ciara. 16 May. 2014 'Reinvention helps Dell steer path through stormy waters'. *Irish Times.*

95. Alfred, Randy. 8 August 2008. 'Aug. 7, 1944: still a few bugs in the system'. *Wired.* archive.wired.com/science/discoveries/news/2008/08/dayintech_0807

96. Irish Times. 11 May 1985. 'Microsoft to set up in Ireland'. *Irish Times.*

97. Irish Times Top 1000 Companies. www.top1000.ie/companies

About the Contributors

Pamela Newenham is a business journalist with the *Irish Times*. She specialises in the areas of innovation, technology, start-ups and entrepreneurship, and was named Technology Reporter of the Year at the 2014 UCD Smurfit School Business Journalist Awards. She has previously written for the *Irish Independent*, the *Irish Examiner*, the *Irish Daily Mail*, *Food & Wine* magazine and NUI Maynooth's *ReSearch* magazine, among other publications. She has a degree in law and a master's degree in journalism.

Joanna Roberts is a freelance journalist who writes about science, business and careers. She is deputy editor of *Horizon*, the European Union's science magazine, and her work has been published in the *Irish Times*, the *Guardian*, *Irish Daily Mail* and *Women Mean Business*. She has a master's degree in international studies and was international manager of the charity Ethiopiad before training as a journalist in London. She splits her time between Dublin and Brussels.

J.J. Worrall is a Dublin-based journalist with digital news agency, Storyful. He has worked extensively as a technology writer for both the *Irish Times* and the *Sunday Business Post*, as well as a number of other titles, including the *Irish Independent* and *TechPro* magazine. He has also written for *The Big Issue* and the music site *State.ie*.

Elaine Burke is a journalist with *Siliconrepublic.com*, covering new media, gadgets, start-ups and tech news. Graduating with a degree in communication studies from Dublin City University in 2007, she went on to work as a writer and editor. *Maternity*, a publication she edited, won Annual of the Year in the 2011 Irish Magazine Awards. In 2013, she was named Tech Writer of the Year at the inaugural Journalism and Media Awards.

Philip Connolly is a journalist with the *Sunday Business Post*, writing about business, legal affairs and technology. A native of Dublin, he studied at University College Dublin and has previously worked for the *Sunday Times* and Amnesty International. He was named Breakthrough Journalist of the Year at the 2013 UCD Smurfit School Business Journalist Awards.

Emmet Ryan is editor of the *Sunday Business Post*'s *Connected* magazine. He has more than a decade of experience covering technology and business and was named Net Visionary Journalist of the Year by the Irish Internet Association in 2009. He once destroyed a laptop with a cappuccino and posted the video on YouTube, but it was for work so that's OK.

Ciara O'Brien has been a technology writer since 2002 and joined the *Irish Times* in 2008. As well as being a business journalist, she is also as part-time gadget geek, and writes on everything from mobile phones and tablets to very expensive TVs and watches. She regularly appears on radio shows trying to convince people she's not anti-Apple, then spends the rest of the day defending herself from blog commenters claiming she's too pro-Apple.